SpringerBriefs in Earth System Sciences

Series Editors

Gerrit Lohmann, Universität Bremen, Bremen, Germany

Justus Notholt, Institute of Environmental Physics, University of Bremen, Bremen, Germany

Jorge Rabassa, Labaratorio de Geomorfología y Cuaternar, CADIC-CONICET, Ushuaia, Argentina

Vikram Unnithan, Department of Earth and Space Sciences, Jacobs University Bremen, Bremen, Germany

SpringerBriefs in Earth System Sciences present concise summaries of cutting-edge research and practical applications. The series focuses on interdisciplinary research linking the lithosphere, atmosphere, biosphere, cryosphere, and hydrosphere building the system earth. It publishes peer-reviewed monographs under the editorial supervision of an international advisory board with the aim to publish 8 to 12 weeks after acceptance. Featuring compact volumes of 50 to 125 pages (approx. 20,000—70,000 words), the series covers a range of content from professional to academic such as:

- A timely reports of state-of-the art analytical techniques
- bridges between new research results
- snapshots of hot and/or emerging topics
- literature reviews
- in-depth case studies

Briefs are published as part of Springer's eBook collection, with millions of users worldwide. In addition, Briefs are available for individual print and electronic purchase. Briefs are characterized by fast, global electronic dissemination, standard publishing contracts, easy-to-use manuscript preparation and formatting guidelines, and expedited production schedules.

Both solicited and unsolicited manuscripts are considered for publication in this series.

Yves Guglielmi

A Review of CO$_2$ Storage Integrity and Fault Zone Risk

 Springer

Yves Guglielmi
Earth and Environment Sciences Area
Lawrence Berkeley National Laboratory
Berkeley, CA, USA

ISSN 2191-589X   ISSN 2191-5903 (electronic)
SpringerBriefs in Earth System Sciences
ISBN 978-3-031-81528-7   ISBN 978-3-031-81529-4 (eBook)
https://doi.org/10.1007/978-3-031-81529-4

This Springer imprint is published by the registered company Springer Nature Switzerland AG
The registered company address is: Gewerbestrasse 11, 6330 Cham, Switzerland

If disposing of this product, please recycle the paper.

Preface

With the growing concern in climate change, governments and industries are intensifying their efforts to plan excess CO_2 storage in geological traps located at several kilometers below the surface. A successful implementation of CCS relies on storing a large amount of CO_2 in multiple storage sites while maintaining storage safe over the long term. Geological faults that are ubiquitous throughout the earth's crust display an extremely complex and large variety of structures and evolutions through time. It is reasonable to imagine that, given the scale of the future CO_2 storage efforts, among a majority of silent faults some will rupture, causing earthquakes and CO_2 flow by-pass the geological traps.

Fault rupture requires sophisticated concepts that were developed from different research disciplines such as Hydrogeology, Mechanics and Seismology. In addition, it appears that several concepts must be combined with each other to best figure the multiple interacting hydromechanical and chemical processes of fault rupture. Beyond the concepts, fault rupture spans over a broad range of time scales, from a few seconds for an earthquake to several days and years for a slow rupture.

All this complexity in the fault physics makes it difficult for the choices to simplify the problem of estimating the fault risk associated to CO_2 storage. The first motivation of this book is to put together key research results on fault mechanics, hydrogeology and induced seismicity that are relevant to CO_2 storage. The second motivation is to provide in the same book a detailed enough physics of fault leakage and induced seismicity, by confronting relevant theories on fault rupture with experimental results from laboratory scale and from mesoscale field experiments.

For the above reasons, we expect this book to be considered as useful by students, scientists and engineers in their attempts to better consider fault zones at CO_2 storage scales.

Berkeley, USA   Yves Guglielmi

The original version of the book has been revised. A correction to this book can be found at
https://doi.org/10.1007/978-3-031-81529-4_6

Acknowledgments

The author would like to thank the following reviewers of this book for their fruitfull comments:

Jens Birkholzer, Lawrence Berkeley National Laboratory, Berkeley, CA 94720, USA

Jonny Rutqvist, Lawrence Berkeley National Laboratory, Berkeley, CA 94720, USA

Frederic Cappa, Géoazur (UMR 7329), University of Nice Sophia-Antipolis, CNRS, IRD, Côte d'Azur Observatory, 06560 Sophia-Antipolis, France

Utkarsh Mital, Lawrence Berkeley National Laboratory, Berkeley, CA 94720, USA

The author would also like to thank the following LBNL colleagues for the discussions that helped him a lot in the writing of this book: Abdullah Cihan, Stanislav Glubokovskikh, Meng Cao, Matthew Reagan and Preston Jordan

The author would like to thank the funding provided by the U.S. Department of Energy under contract FP00015629, titled 'Managing a Gigatonne CCS Future: A Framework for Basin-Scale Storage Optimization Based on Geomechanical Studies,' and under Proposal Number FP00014702, titled 'National Risk Assessment Partnership (NRAP) Phase III'.

Introduction

After decades of research on carbon capture and storage (CCS), the world needs to finally move from pilot tests and demonstration experiments to industrial-scale implementation (Birkholzer et al., 2015; Shu et al., 2023). In different countries, CCS deployment over the coming years and decades will likely create multiple large storage projects (or clusters of integrated projects, or hubs) across selected sedimentary basins. These projects will involve the injection of substantial volumes of CO_2, leading to large-scale pressure increases, accompanied by associated stress perturbations within the subsurface (Nicot, 2008; Birkholzer and Zhou, 2009; Zhou et al., 2010). These activities may potentially cause unwanted geomechanical effects, such as the formation of leakage pathways in the caprock or the seismic reactivation of critically stressed faults (Rutqvist et al., 2012; Rutqvist et al., 2016; Zoback and Gorelick, 2012; Ellsworth, 2013; Vilarrasa et al., 2019).

Fault reactivation can be a concern even if the large-scale pressure changes are smaller than the pressure thresholds (i.e., fracturing pressure) typically employed to ensure that local injection pressures do not cause geomechanical damage in the reservoir and the confining units (Rutqvist et al., 2007). It follows that large-scale pressure buildup can be a limiting factor for sequestration capacity (Thibeau et al., 2014), because the possibility of pressure-related impacts of individual projects needs to be considered. Furthermore, in a future world with CCS being a fully deployed technology, sedimentary basins with interconnected reservoirs might host multiple storage sites between which pressure interference can be expected (Birkholzer and Zhou, 2009; Zhou et al., 2010), potentially adding a further concern for storage security and further constraints on storage capacity.

In this review, we explore how the complexity of the rheological heterogeneity of faults is considered in generating permeable pathways for brine or CO_2 to flow out of the storage reservoir along the faults. We focus on the potential for flow paths to grow along fault zones and trigger the loss of integrity of caprocks sealing the storage reservoir. We define this as the fault leakage risk. We also explore how building pore pressure while storing supercritical CO_2 into a reservoir layer can generate seismicity around and far deeper from the storage volume in the basement of the sedimentary basin. We define this as the seismic risk induced by CO_2 storage. We focus on fault zones affecting sedimentary basin and their upper basement limit, i.e., the behavior of faulted layered lithostratigraphic systems between the surface and perhaps 6–8km depth. We first provide an overview of the current workflows and concepts used to estimate both leakage risk through faults across caprocks and changes in seismicity rates caused by fluid injections (Chap. 1). We discuss how these workflows integrate simplified fault zones physics based on clay content, friction and stress or stress rates to describe the seismic activation and leakage of faults.

In Chap. 2, we go deeper into the geomechanics of basin faults and on the implications for CO_2 storage at scale. One key question is the potential effect of brittleness and of brittle-ductile limits on faults within basins. We discuss how the presence of faults within clay-rich caprocks affects their frictional dilatant properties, which in turn alters their leakage and seismogenic potentials. We furthermore explore various constitutive relationships for the plastic deformation of faults and how to relate contraction/dilation with softening/hardening leading to stable/unstable fault slip. Specifically, we propose that a Cam-Clay model may provide a more general mechanical framework compared to a Coulomb model for describing a large variability of fault activation scenarios.

Chapter 3 is a review of all factors influencing permeability change at fault rupture. This chapter is focused on the complex coupling between fault mechanics and permeability. We first review different laboratory-scale experiments and compare these to fault permeability measurements from field-scale experiments. We observe that fault permeability tends to drastically decrease with shear at laboratory scale while a much smaller-to-no decrease is observed in the field. We finally propose a new conceptual model for the fault permeability evolution with slip and discuss how it can be considered in constitutive relationships for the coupled hydromechanical behavior of faults.

Chapter 4 is dedicated to induced seismicity. It appears that induced seismicity is currently modeled using one dominant approach based on the physics of rate-and-state friction (Dieterich, 1972, 1979; Marone, 1998). We review the key hypotheses and concepts of this theory and describe how it has been applied to model changes in natural seismicity rates related to seismicity rates due to CO_2 injections at basin scale. Then, we compare this physics based on friction with another model one based on plastic instability, referred to as Cam-Clay model. This plastic approach is relatively new and has never really been tested for basin scale injections. Based on an extension of the Cam-Clay model to dynamic processes, we particularly explore how slow slip initiated on a fault away from a CO_2 storage project can accelerate with time into an earthquake rupture. We use existing codes from the literature to estimate if this is possible when considering a basin fault initially mechanically stable and relatively far from critical state of stress.

We conclude this review by asking what the hydromechanical behavior of faults as discussed above means for carbon storage management at basin scale. We propose a general framework for assessing these impacts which we hope will help unify leakage and induce seismicity workflows in the future.

References

Birkholzer, J. T., & Zhou, Q. (2009). Basin-scale hydrogeologic impacts of CO_2 storage: Capacity and regulatory implications. *International Journal of Greenhouse Gas Control, 3*(6), 745–756.

Birkholzer, J. T., Oldenburg, C., & Zhou, Q. (2015). CO_2 migration and pressure evolution in deep saline aquifers. *International Journal of Greenhouse Gas Control, 40*, 203–220. ISSN 1750-5836. https://doi.org/10.1016/j.ijggc.2015.03.022

Dieterich, J. H. (1972). Time-dependent friction in rocks. *Journal of Geophysics Research, 77*(20), 3690–3697.

Dieterich, J. H. (1979). Modeling or rock friction, 1, Experimental results and constitutive equations, *Journal of Geophysics Research, 84*, 2161–2168.

Ellsworth, W. L. (2013). Injection-induced earthquakes. *Science, 341*, 6142. https://doi.org/10.1126/science.1225942

Marone, C. (1998). Laboratory-derived friction laws and their application to seismic faulting. *Annual Review of Earth Planetary Science, 26*(1), 643–696.

Nicot, J.-P., Oldenburg, C. M., Bryant, S. L., & Hovorka, S. D. (2008). Pressure perturbations from geologic carbon sequestration: Area-of-review boundaries and borehole leakage driving forces: Presented at the 9th International Conference on Greenhouse Gas Control Technologies (GHGT-9), Washington, D.C., 16–20 Nov 2008. GCCC Digital Publication Series #08-03h.

Rutqvist, J. (2012). The geomechanics of CO_2 storage in deep sedimentary formations. *International Journal of Geotechnical and Geological Engineering, 30*, 525–551.

Rutqvist, J., & Tsang C.-F. (2002). A study of caprock hydromechanical changes associated with CO_2 injection into a brine aquifer. *Environmental Geology, 42*, 296–305.

Rutqvist, J., Birkholzer, J., Cappa, F., & Tsang C.-F. (2007). Estimating maximum sustainable injection pressure during geological sequestration of CO_2 using coupled fluid flow and geomechanical fault-slip analysis. *Energy Conversion and Management, 48*, 1798–1807. https://doi.org/10.1016/j.enconman.2007.01.021

Rutqvist, J., Rinaldi, A. P., Cappa, F., Jeanne, P., Mazzoldi, A., Urpi, L., Guglielmi, Y., & Vilarrasa, V. (2016). Fault activation and induced seismicity in geological carbon storage—Lessons learned from recent modeling studies. *Journal of Rock Mechanics and Geotechnical Engineering, 8*(6), 789–804. https://doi.org/10.1016/j.jrmge.2016.09.001

Thibeau, S., Bachu, S., Birkholzer, J., Holloway, S., Neele, P., & Zhou, Q. (2014). Using pressure and volumetric approaches to estimate CO_2 storage capacity in deep saline aquifers. *Energy Procedia, 63*, 5294–5304.

Shu, D. Y., Deutz, S., Winter, B. A., Baumgärtner, N., Leenders, L., Bardow, A. (2023 May). The role of carbon capture and storage to achieve net-zero energy systems: Trade-offs between economics and the environment. *Renewable and Sustainable Energy Reviews, 178*, 113246.

Vilarrasa, V., Carrera, J., Olivella, S., Rutqvist, J., & Laloui, L. (2019). Induced seismicity in geologic carbon storage. *Solid Earth, 10*, 871–892. https://doi.org/10.5194/se-10-871-2019

Zhou, Q., Birkholzer, J. T., Mehnert, E., Lin, Y.-F., & Zhang, K. (2010). Modeling basin- and plume-scale processes of CO_2 storage for full-scale deployment. *Groundwater, 48*(4), 494–514.

Zoback, M. D., & Gorelick, S. M. (2012). Earthquake triggering and large-scale geologic storage of carbon dioxide. *Proceedings of the National Academy Sciences, 109*, 10164–10168.

Contents

Chapter 1
Current Fault Seal and Induced Seismicity Workflows

Abstract In this chapter, we give an overview of some of the most commonly used techniques for risk assessment of fault sealing and for induced seismicity. Fault sealing has long been a concern in reservoir/caprock engineering (Knipe in Structural and tectonic modelling and its application to petroleum geology. Elsevier, Stavanger, 1992; Ingram et al. in Hydrocarbon seals: Importance for exploration and production. Elsevier, Singapore, 1997). We first present two complementary methods used to estimate the across-fault and along-fault sealing integrity, respectively. The first method considers the fault as a thin membrane where permeability depends on the membrane clay mineral content. The method was develop from fault-seal observations in clastic sedimentary sequences under a low differential stress (Knipe in Structural and tectonic modelling and its application to petroleum geology. Elsevier, Stavanger, 1992 for ex.). Empirical relationships are made between the clay mineral content of the sedimentary layers affected by the fault, the fault offset and the fault permeability (deduced from the history matching of pore pressure signals from both fault compartments). The second method considers the fault as a frictional interface, and it calculates the ratio between the stresses normal and tangential to the interface knowing the regional stress regime. This method was developed to describe fault permeability variations in low permeable rocks subject to deep and high differential stresses (Ferrill et al. in Solid Earth 11:899–908, 2020). It relies on observations that fault permeability vary at fault mechanical activation through mechanisms of slip and dilation. The method uses the ratio between stresses to conduct a qualitative estimate of how much the fault figured as an interface tends to dilate under slip, and considers that the higher the tendency to dilate the higher the along-fault permeability should be without giving any quantitative estimate. Second, we discuss how seismicity induced by CO_2 storage projects can be considered. Compared to fault seal this is a more recent concern that attracted more attention when seismicity started to be induced by wastewater injections or shale gas hydraulic fracturing, and came the need to better assess the seismic risk related to large scale CO_2 injections (Zoback and Gorelick in Proceed Nat Acad Sci 109:10,164–10,168, 2012). Here, we introduce some of the fault physics used in estimating induced seismicity, and how it can be applied to CO_2 projects induced seismicity rick assessment. Finally, this first chapter allows us to highlight the complex and multiple fault zone attributes that need to be considered at basin scale to assess fault seal and seismic risk.

Keywords Workflows · Shale gouge ratio · Shale smear factor · Slip tendency · Critical pressure theory · Rate-and-state friction · Clay content · Fault thickness · Fault throw

1.1 Static Estimation of Fault Sealing Properties Based on Fault Clay Content and Stress

- Influence of clay content

The clay content in fault rocks is widely recognized as a key factor influencing fault hydraulic and frictional properties (Ashman & Faulkner, 2023; Crawford et al., 2008; Ikari et al., 2007; Kohli & Zoback, 2013; Pei et al., 2015; Ruggieri et al., 2021; Takahashi et al., 2007; Zhang et al., 2020). Based on the percentage of clay minerals, Pei et al. (2015) and, more recently, Ashman and Faulkner (2023) suggest that there is a transition in the hydromechanical response of fault rocks with more than 40% clay mineral content. Fault rocks with more than 40% clay minerals may be called shale smears, corresponding to fault zones with a low permeability and low friction. In a sedimentary basin, the fault clay content depends on the mechanical abrasion and other types of brittle-to-ductile deformation of the layers of the lithostratigraphic series (Giger et al., 2013; Yielding et al., 1997).

One common approach routinely used in the oil and gas industry to estimate fault sealing properties is to deploy empirical relationships between the clay mineral content of the fault material and the fault permeability. The clay mineral content is then defined by simple parameters, most commonly the shale gouge ratio (SGR) which gives an estimate of the potential clay content of the fault zone (Yielding et al., 1997) and the shale smear factor (SSF) which gives an estimate of the extent of the shale smear in the fault zone (Lindsay et al., 1993 and Fig. 1.1). In the simple case shown in Fig. 1.1b, the SGR and SSF are calculated using Eqs. 1.1 and 1.2:

$$SGR = \frac{\sum(\text{Volumetric clay fraction in layer } i \times \text{thickness layer } i)}{\text{throw}} \times 100\%$$

$$= \frac{\sum(V_{cli} \times H_i)}{\text{throw}} \times 100\%, \tag{1.1}$$

$$SSR = \frac{\text{throw}}{\text{shale layer thickness}}. \tag{1.2}$$

Empirical relationships relate the fault clay content to the fault throw and to the cumulative thickness of the shale layers intersecting the fault. These relationships are proxies meant to quantify the amount of clay smear occurrence along a fault

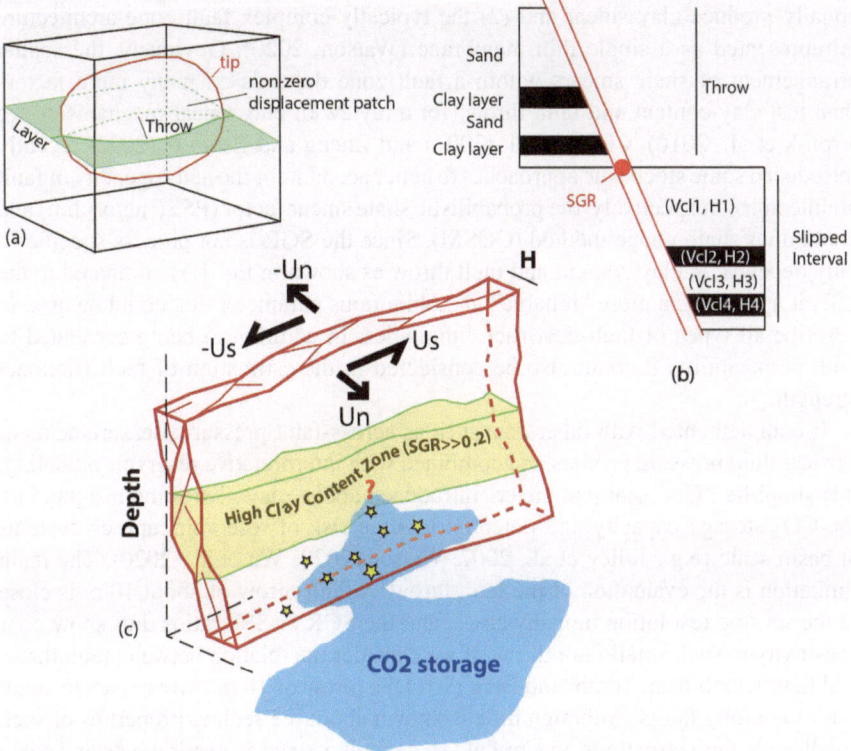

Fig. 1.1 Conceptual view of a fault zone affecting a sedimentary sequence. **a** Main fault attributes usually considered in leakage workflows. **b** Example of shale gouge evolution calculated from the SGR formula. **c** Detailed view of a fault zone with a thickness H, asperities and clay content "layering". U_n and U_s are displacements normal and parallel to the fault zone. Yellow stars figure induced earthquakes

surface. The correlation between the SGR and SSF and the fault permeability is then established from laboratory measurements on cores, extrapolated to an estimated fault zone thickness. The fault zone thickness can be roughly correlated to the fault throw, and thus to the SGR and SSF. The advantages of such a fault seal potential approach are that it relies on a relatively limited number of parameters that can be measured in the field. Indeed, the clay content of the sedimentary layers can be deduced from gamma ray borehole logs and the fault throw can be estimated from the offset of seismic reflection interfaces. Thus, a mapping of the clay content on the fault surface can be done (Fig. 1.1c, d). Areas of low clay content (SGR < 15%) and of low shale smear extent (SSF > 7) are considered potentially permeable, whereas areas of high clay content (SGR > 40%) and low shale smear factor (SSF < 4) are considered potentially sealing.

Due to its simplicity, the SGR/SSF approach tends to overestimate the amount of clay in fault rocks because (1) it considers that all shale layers affected by a fault equally produce clay smear and (2) the typically complex fault zone architecture is represented as a single thin membrane (Watson, 2020). Obviously, the spatial arrangement of shale smears within a fault zone depends on many more factors than just clay content and fault throw (for a review of clay smear mechanisms see Vrolijk et al., 2016). Childs et al. (2007) and Zheng and Espinoz (2022) recently introduced some stochastic approaches to better account for the heterogeneity of fault architectures, respectively, the probabilistic shale smear factor (PSSF algorithm) and continuous shale gouge method (CGSM). Since the SGR is not process-specific (it only depends on clay content and fault throw as shown in Eq. 1.1) compared to the SSF, it appears as a more "reliable" and ubiquitous parameter that could be used to describe all types of fault-host rock lithologies. In addition to being correlated to fault permeability, it could also be considered in the estimation of fault frictional strength.

If complemented with other data such as across-fault pressure measurements or vertical fluid pressure profiles and combined with interpretative reservoir modeling, this simplified fault seal parameters introduced above may give some insights into the CO_2 storage capacity and potential leakage risk of reservoir/caprock systems at basin scale (e.g., Jolley et al., 2007; Watson, 2020; Wu et al., 2020). The main limitation is the evaluation of the fault throw. A fault throw of about 10 m is close to the seismic resolution in many cases, and the SGR or SSF ratios thus show poor sensitivity to such small fault throw. If we consider the relation between fault throw and fault length from Torabi and Berg (2011), a throw of 10 m corresponds to small 1-to-3 km long faults. Although little is known about the sealing properties of such small-scale fault structures, in situ field scale fault activation experiments and some outcrop analog studies suggest that subseismic scale faults may become permeable under a reduced effective stress (Guglielmi et al., 2020; Manzocchi et al., 2010). Thus applying SGR and SSF algorithms to structures mapped through seismic surveys may lead to predictions that either over or underestimate the degree of fault sealing.

- Influence of the regional stress

The SGR mapping on a given fault surface can be compared to an estimation of the slip tendency ST of a fault. This parameter is defined by the equation $ST = \tau/\sigma_n$ with τ and σ_n respectively being the shear and normal stresses applied on the fault (Ferrill et al., 2020; Morris et al., 1996). For a given fault surface with waviness and asperities at various scales, ST will vary as a function of the local fault surface orientation relative to the local stress orientation, defining areas of the fault surface that have a significantly higher tendency to slip than others. ST can be complemented by another parameter, the fault dilation tendency defined as $DT = (\sigma_1 - \sigma_n)/(\sigma_1 - \sigma_3)$. DT is particularly relevant in normal and strike-slip stress regimes where the horizontal minimum principal stress is perpendicular-to-oblique to the fault surface and thus will favor opening of the fault. Compared to the static SGR mapping, ST and DT allow defining zones of potential rupture and permeability generation along the fault zone. In the case of a high ST-SD, dilation induced by slip will be the phenomenon driving fault opening.

Ward et al. (2016) found that vertical patches of high slip tendency along single faults in the Southern North Sea were close to vertical fluid flow paths detected by seismic characterization. Miocic et al. (2020) found that calculated high slip tendency patches on the Coyote Wash Fault crossing the St. Johns dome (Colorado) coincided with travertine deposits along the fault trace at the ground surface. Nevertheless, in the same Colorado area but on another fault system, Naruk et al. (2019) observed that fault leakage as evidenced by the occurrence of travertine deposits along the fault surface trace can also occur in areas of low ST if fluid pressures applied on the fault can get close to the normal stress and thus reduced effective normal stress to close to zero. This may happen in the case of high reservoir overpressures, and can be described by the fracture gradient for example defined as FG = $((\mu/1 - \mu) \times$ (OBG − Pp)) + Pp where μ is the Poisson's ratio, OBG the overburden stress and Pp the pore fluid pressure (Eaton, 1969).

The effects of both the optimal fault orientation to slip and of elevated fluid pressures on fault dilatant rupture and potential leakage were described in Zoback (2007). In another paper, Wiprut and Zoback (2002) highlighted that both these conditions may be met in cases of reservoir-bounding faults when the across-fault pore pressure differential can be significant. Recent studies show that such conditions can also exist in compartmentalized reservoirs where the fault permeability is lower than the reservoir permeability (Wu et al., 2020), and can be aggravated by reservoir depletion through complex poroelastic stressing related to differential compaction (Hettema, 2020).

Contemporary workflows to estimate fault permeability and leakage remain qualitative as they highlight fault areas where leakage could potentially occur mainly because of low clay content and high tendency to slip and dilate, but they do not give an estimate of the characteristics of the leakage pathway, such as its permeability and size. More complicated workflows have recently been developed and successfully calibrated against natural CO_2 seapage along a major fault zone in Utah (Snippe et al., 2022). In that case, the workflow utilizes a numerical flow/mechanical model (MoReS, Regtien et al., 1995) considering a simplified geological model of the fault zone with fault relative permeability to CO_2 phases, capillary and pore pressure as well as a stress-dependent permeability law. Application of such simulators remain relatively rare at basin scale.

1.2 Induced Seismicity Workflows Based on Geomechanics and on Fault Frictional Physics

There are two main scenarios currently considered in the risk of induced seismicity related to fault reactivation by CO_2 storage (Cheng et al., 2023 for a review). The storage site is considered to be set away from the known "major" faults of the basin, but it cannot be avoided that some unknown faults be relatively close to the site. Then, a first scenario is that pressure builds up directly in a fault zone hydraulically connected to the storage reservoir which changes the effective normal stress and may trigger an earthquake. This scenario may be complicated by the possibility that the supercritical CO_2 also enters the fault which may modify the seismic stability of the fault and/or produce CO_2 leakage along and up the fault. The other scenario is that faults located below the storage reservoir, at the limit between the basement and the sedimentary layers, are activated by the poroelastic transfer of stress change induced by pressure build up in the reservoir layer.

To describe these scenarios, there are mainly two existing physics, the critical pressure theory (CPT, Shapiro, 2015) and the rate-and-state friction (RST, Dieterich, 1972). For a comparison of these two theories refer to Wenzel (2017). Note that the CPT originated from observations that the spatio-temporal distribution of seismicity induced by fluid injections in a deep borehole could be explained by the hydraulic diffusion of fluid pressure away from the borehole (Shapiro et al., 1997). The RST was developed from laboratory studies of slip on faults represented as frictional interfaces and initially applied to describe natural seismicity on deep faults at seismogenic depth (Dieterich, 1978a). Later, the theory was "exported" to describe changes in seismicity rates including changes caused by fluid injections in reservoirs (Chang & Segall, 2016; Dieterich et al., 2015). This short history of these two theories explains most of the differences in the concepts used to model seismicity induced by fluid storage at basin scale. Note that these are the two dominant theories and the authors of this paper do not intend to make an exhaustive review of all other existing theories. For such a review refer to Gaucher et al. (2015).

In the CPT theory, the number of induced seismic events is proportional to the rate of pressure change, and unrelated to the tectonic background seismicity as well as fault strength weakening related to changes in friction and cohesion with strain. Only changes in effective normal stress related to an increase in fault pore pressure are considered. Thus faults are initially stable, but the ones that rupture will be close to criticality, and as soon as Coulomb failure occurs the rupture will be considered seismic. This means that the number of seismic events is proportional to pressure variation with time, within a statistical network of fractures. This approach works relatively well when describing induced seismicity relatively close to the injection source. Nevertheless, recent studies show that even close to injection there can be significant aseismic slip on activated faults and this aseismic slip may also be an important driver of induced seismicity at locations where there is no fluid pressure diffusion (Cappa et al., 2019; Cebry et al., 2022; DeBarros et al., 2018; Duboeuf et al., 2017; Eyre et al., 2019; Guglielmi et al., 2015; Pepin et al., 2022; Wynants-Morel et al., 2020). These observations show for example that some seismic events can be triggered at distances that are larger than how far fluid pressure diffuses and/or at distances too large compared to the ones estimated at a given hydraulic diffusion time. For the CPT to be applied to the scenario of faults that are not connected to the injection source, it would still require an infinitesimal pore pressure change to activate these faults. Such an assumption figures that these faults are initially very close to critical state of stress and that when activated they spontaneously trigger an earthquake.

In the RST theory, the change in the seismicity rate induced by fluid injection is related to the background natural seismicity and seismicity rate, a characteristic time decay which depends on a fault friction parameter "a", on the temporal evolution of the normal effective stress, and the Coulomb stressing rate (Dieterich, 1994; Dieterich et al., 2015). In addition, a complex fault friction response is considered allowing for shear stress to vary upon injection just as the effective normal stress varies. The complex friction response allows to condition the seismic instability of the fault to a fault frictional decrease with changes in fault slip velocity and the state of fault asperities. In this theory, not all faults may produce earthquakes at Coulomb failure, and if they do, seismicity only occurs after some time delay. The issue with this theory is that faults must initially be at rupture, which may be "tricky" to consider in a geomechanical simulation model where a so-called consolidation phase is usually necessary before loading the model with a fluid injection. In recent works, this difficulty is overcome by running a geomechanical model to calculate changes in Coulomb stress rates ΔCSR and using a separate RST approach to relate the ΔCSR with the seismicity rates (Candela et al., 2019; Luu et al., 2022). The RST has also some limitation when applied to intraplate basin injections where there is very limited natural seismicity before the injections start (see Chap. 4 of this paper for more details).

To summarize, the CPT approach may be suitable to describe seismicity on faults hydraulically connected to the injection source. The approach has the advantage faults are not required to be at rupture before injection starts as is in the RST approach. The CPT can also be "sufficient" in cases with no tectonic background activity. The RST approach, on the other hand, may better represent how a poroelastic stress change can accelerate slip rates on initially slowly slipping faults rooted in the basement but disconnected from the storage reservoir. The RST approach is much more complex and harder to apply than the CPT. It may work efficiently to describe fluid injection in tectonically active basins. In the CPT, the assumption that earthquakes trigger spontaneously at Coulomb failure may lead to overestimations of the effects of fluid pressure increases on induced seismicity and to the underestimation of other processes such as aseismic deformation. In the RST, the a and b parameters initial values precondition faults for seismic rupture (a rupture being defined as seismic if $a-b < 0$, see details in Chap. 4) although their tuning better describes changes in seismicity rates with time.

This quick overview highlights that there is no current workflow that fully represents to the large diversity of seismic fault behaviors and their effects on induced earthquakes. For example, the two fault physics models described above strongly underestimate (or do not even consider) the influence of aseismic fault creep on induced seismicity. The two models are also mostly based on the Coulomb failure of faults which limits the consideration of absolute stresses in fault rupture (for example, the role of the intermediate principal stress in fault rupture is not considered in Coulomb failure). Another point that is barely considered in induced seismicity prediction is the effect of fluid compressibility or viscosity changes that are likely to be relevant when a supercritical CO_2 fluid enters a fault zone.

1.3   Main Fault Characteristics to Consider at Basin Scale

The main question when considering faults at basin scale is how to account for the strong variability of their geological properties that result from the complex geodynamic history of the basin. Thanks to seismic reflection methods and comparisons with geological outcrop studies, some key fault characteristics can be defined, such as the fault surface aspect ratio (which is the ratio between fault length along strike versus along dip), the fault thickness that includes fault zone internal complexities such as core and damage zone, the throw that varies from zero at fault tip to a maximum value at the center of the fault surface, and the fault orientation (dip angle and dip direction). In Sect. 1.1, we described relationships between the fault throw and the clay content as described by the shale gouge ratio which gives a proxy to the across-fault sealing capacity with consequences on the reservoir compartmentalization and thus storage capacity. We also described how the orientation of the fault versus the orientation and magnitude of the stress tensor is used to estimate the potential for slip and dilation of a fault. Section 1.2 highlights that there is a need for more detailed fault zone representation to match hydromechanical laws that can help assess leakage and induced seismicity. Here, structural geologists suggest many empirical correlations between some of the fault characteristics listed above. These correlations take the form of power laws that indicate fractal dimensions can be used to approximately describe the geometric attributes of fault populations (Yielding et al., 1992). Although these laws depend on the complex current/past geodynamical context and on the basin lithostratigraphy, we shortly describe how some of these empirical rules may inform workflows dedicated to fault leakage and induced seismicity.

- Current fault statistics and empirical relationships between fault attributes

A first power law relationship relates fault thickness H and the maximum fault throw T in the form of $H = aT^b$, where a and b are constants (Torabi & Berg, 2011). However, this is no general relationship and the data are often quite disperse. The strong dispersion in the data mainly comes from the fact that many faults have experienced several tectonic phases of reactivations that resulted in highly nonlinear changes in the damage zone thickness. Nevertheless, by compiling a limited amount of data from Schueller et al. (2013), Childs et al. (2009), Bense et al. (2013), Knott et al. (1996), Mitchell and Faulkner (2009) and Shipton and Cowie (2001), Houwers et al. (2015) highlight that values for a and b range between 0.1-to-12 and 0.25-to-0.67, respectively. They show that a smaller a value relates to the faulted host rock clay content, or in other words that, for the same throw, clay faults are statistically thinner than other fault lithologies. A second power law relationship exists between the maximum fault throw and the along-strike fault length L. This relationship mainly works in the case of normal faults. It is in the form of $T = 10^c L$ with c varying between -3 and 0 (Kim & Sanderson, 2005). Schematically, this correlation says that longer faults have larger throws. The third power law describes the cumulated frequency of faults of a given length in each basin, highlighting the fractal nature of fault populations (Yielding et al., 1992). Examples from different basins show that 60–70% of faults have a throw < 50 m at basin scale and that less than 5–10% of faults display throws > 300 m. Other statistics are proposed by structural geologists that in general highlight two key parameters that explain fault geometry, respectively, (1) the heterogeneity of the basin lithostratigraphic sequence that induces "mechanical layering" which means faults preferentially develop in some sedimentary layers at given basin depths and (2) the presence of pre-existing structures which for example can lead to the development of more recent faults in the sedimentary basin above an older fault affecting the deep basement (see some examples in Duchek et al., 2004; Ishii, 2016; Langhi & Borel, 2008; Lee et al., 2018; Redpath et al., 2022; Roche et al., 2020, 2021).

- Hydromechanical consequences

Following the statistics described above, a large-scale fault will display a larger throw and a larger thickness than a short fault. There is large body of field observations that show that large-scale faults are preferential flow paths for fluids. First, they offer a large volume for flow paths to develop, and second, their large length favors along-fault leakage between the multiple reservoirs that they connect. For example, Vadakkepuliyambatta et al. (2013) observe that the location of gas chimneys in the SW Barent Sea matches with the presence of major faults deeply rooted into the basement. Zhao et al. (2023) suggest that faults with a throw greater than 200 m have the greatest influence in draining shale gas of the Sichuan Basin (China).

The large throw can also have key consequences on the way fluids can escape from the fault zone. A large throw results in a large reservoir offset. For example, a 300m throw can fully offset a 200 m-thick reservoir. Thus, the reservoir on one side of the fault can be in contact with a caprock layer on the other side. The fault is then "bounding" the reservoir. Ostanin et al. (2017) observe that leakage along such fault-bounding reservoirs is the dominant mechanism driving hydrocarbon leakage observed in Barents Sea basins (Norway). Wiprut and Zoback (2002) estimate that in addition to stress and stress variations, locally elevated pore pressure preferentially develops in reservoirs bounded by large fault offsets in the North Sea. Tryon et al. (2012) observe that the large Marmara Basin (Turkey) bounding fault and other tectonically active faults are the main flow paths that expel basin fluids. Nevertheless, Sect. 2.1 shows that the dominant role of large faults in fluid leakage must be moderated because these faults are highly complex features. Given their length and large throw, some of these large fault zones may display larger clay content if they cut a clay-rich lithostratigraphic series (see discussion about a correlation between throw and shale gouge ratio) and thus be relatively low permeable. From a mechanical point of view, these faults may have weaker strength (or are closer to their residual strength) than smaller ones if we roughly consider that a thicker damage zone is a proxy to a larger plastic deformation during faulting. They must then be easier to reactivate. If in addition they are pressurized by leaking fluids (even partially), they can be a source of relatively larger induced earthquakes compared to small faults as calculated by Zhang et al. (2013 and see more details in Sect. 4.2).

Given their statistically high number and multiple intersections, small-scale faults typically have more tortuous flow patterns at reservoir scale and/or may compartmentalize the reservoir, significantly reducing its storage capacity and favoring local unexpectedly high pressure build up (Zhou et al., 2008). The statistical laws described before show that small-scale faults may explain 40-to-55% of the total strain in a basin (Walsh et al., 1991). Thus, their role in aseismic slow background deformations may be significant although there is little data on this topic. This means that a basin volume with a high density of small faults may display a strong stress heterogeneity that can favor local fault reactivation even when changes in fluid pressure are small. Guglielmi et al. (2020 and see details in Chap. 3) show that small faults can display dilatant shear displacements high enough to trigger significant fault permeability changes and moderate induced seismicity. Finally, the so-called "mechanical stratigraphy" describes the fact that small-scale faults often limited to one depth interval within the lithostratigraphic series (Ishii, 2016; Lee et al., 2018; Redpath et al., 2022; Roche et al., 2021), which means that the leakage flow paths along such faults are shorter and there is less concern of a direct leakage connection to the near surface or surface.

References

Ashman, I. R., & Faulkner, D. R. (2023). The effect of clay content on the dilatancy and frictional properties of fault gouge. *Journal of Geophysical Research: Solid Earth, 128*, e2022JB025878. https://doi.org/10.1029/2022JB025878

Bense, V. F., Gleeson, T., Loveless, S. E., Bour, O., & Scibek, J. (2013). Fault zone hydrogeology. *Earth-Science Reviews, 127*, 171–192. https://doi.org/10.1016/j.earscirev.2013.09.008

Candela, T., Osinga, S., Ampuero, J.-P., Wassing, B., Pluymaekers, M., Fokker, P. A., Van Wees, J. D., De Waal, H. A., & Muntendam-Bos, A. G. (2019). Depletion-induced seismicity at the Groningen gas field: Coulomb rate- and -state models including differential compaction effect. *Journal of Geophysical Research: Solid Earth, 124*, 7081–7104

Cappa, F., Scuderi, M. M., Collettini, C., Guglielmi, Y., & Avouac, J. P. (2019). Stabilization of fault slip by fluid injection in the laboratory and in situ. *Science Advances, 5*(3), eaau4065.

Cebry, S. B. L., Ke, C. Y., & McLaskey, G. C. (2022). The role of background stress state in fluid-induced aseismic slip and dynamic rupture on a 3-m laboratory fault. *Journal of Geophysical Research: Solid Earth, 127*(8), e2022JB024371.

Chang, K. W., & Segall, P. (2016). Injection-induced seismicity on basement faults including poroelastic stressing. *Journal of Geophysical Research: Solid Earth, 121*, 2708–2726. https://doi.org/10.1002/2015JB012561

Cheng, Y., Liu, W., Xu, T., Zhang, Y., Zhang, X., Xing, Y., Feng, B. & Xia, Y. (2023). Seismicity induced by geological CO_2 storage: A review. *Earth-Science Reviews, 239*, 104369. ISSN 0012-8252. https://doi.org/10.1016/j.earscirev.2023.104369

Childs, C., Manzocchi, T., Walsh, J. J., Bonson, C. G., Nicol, A., & Schoepfer, M. P. J. (2009). A geometric model of fault zone and fault rock thickness variations. *Journal of Structural Geology, 31*, 117–127.

Childs, C., Walsh, J. J., Manzocchi, T., Strand, J., Nicol, A., Tomasso, M., Schopfer, M. P. J., & Aplin, A. C. (2007). Definition of a fault permeability predictor from outcrop studies of a faulted turbidite sequence, Taranaki, New Zealand. *Special Publication of the Geological Society of London, 292*, 235–258. https://doi.org/10.1144/SP292.14

Crawford, B. R., Faulkner, D. R., & Rutter, E. H. (2008). Strength, porosity, and permeability development during hydrostatic and shear loading of synthetic quartz-clay fault gouge. *Journal of Geophysical Research: Solid Earth, 113*(B3), 1–14.

De Barros, L., Guglielmi, Y., Rivet, D., Cappa, F., & Duboeuf, L. (2018). Seismicity and fault aseismic deformation caused by fluid injection in decametric in-situ experiments. *C. r. Geosci., 350*(8), 464–475.

Dieterich, J. H. (1972). Time-dependent friction in rocks. *Journal of Geophysical Research, 77*(20), 3690–3697.

Dieterich, J. H. (1978a). Preseismic fault slip and earthquake prediction. *Journal of Geophysical Research: Solid Earth, 83*, 3940–3947.

Dieterich, J. (1994). A constitutive law for rate of earthquake production and its application to earthquake clustering. *Journal of Geophysics Research: Solid Earth, 99*(B2), 2601–2618. https://doi.org/10.1029/93jb02581

Dieterich, J. H., Richards-Dinger, K. B., & Kroll, K. (2015). Modeling injection-induced seismicity with the physics-based earthquake simulator RSQSim. *Seismological Research Letters, 86*, 4. https://doi.org/10.1785/0220150057

Duboeuf, L., De Barros, L., Cappa, F., Guglielmi, Y., Deschamps, A., & Seguy, S. (2017). Aseismic motions drive a sparse seismicity during fluid injections into a fractured zone in a carbonate reservoir. *Journal of Geophysical Research: SolidEarth, 122*, 8285–8304. https://doi.org/10.1002/2017JB014535

Duchek, A. B., McBride, J. H., John Nelson, W., & Leetaru, H. E. (2004). The cottage grove fault system (Illinois Basin): Late Paleozoic transpression along a Precambrian crustal boundary. *GSA Bulletin, 116* (11–12): 1465–1484. https://doi.org/10.1130/B25413.1

Eaton, B. A. (1969). Fracture gradient prediction and its application in oilfield operations. *JPT, 21*(10), 25–32. https://doi.org/10.2118/2163-PA

Eyre, T. S., Eaton, D. W., Garagash, D. I., Zecevic, M., Venieri, M., Weir, R., & Lawton, D. C. (2019). The role of aseismic slip in hydraulic fracturing–induced seismicity. *Science Advances, 5*(8), eaav7172.

Ferrill, D. A., Smart, K. J., & Morris, A. P. (2020). Resolved stress analysis, failure mode, and fault-controlled fluid conduits. *Solid Earth, 11*, 899–908. https://doi.org/10.5194/se-11-899-2020

Gaucher, E., Schoenball, M., Heidbach, O., Zang, A., Fokker, P., Van Wees, & J., Kohl, T. (2015). Induced seismicity in geothermal reservoirs: A review of forecasting approaches. *Renewable and Sustainable Energy Reviews, 52*, 1473–1490. https://doi.org/10.1016/j.rser.2015.08.026

Giger, S. B., Clennell, M. B., Çiftçi, N. B., Harbers, C., Clark, P., & Ricchetti, M. (2013). Fault transmissibility in clastic-argillaceous sequences controlled by clay smear evolution. *American Association of Petroleum Geologists Bulletin, 97*(5), 705–731. https://doi.org/10.1306/10161211190

Guglielmi, Y., Cappa, F., Avouac, J. P., Henry, P., & Elsworth, D. (2015). Seismicity triggered by fluid injection–induced aseismic slip. *Science, 348*(6240), 1224–1226.

Guglielmi, Y., Nussbaum, C., Jeanne, P., Rutqvist, J., Cappa, F., & Birkholzer, J. (2020). Complexity of fault rupture and fluid leakage in shale: Insights from a controlled fault activation experiment. *Journal of Geophysical Research: Solid Earth, 125*, e2019JB017781. https://doi.org/10.1029/2019JB017781

Hettema, M. (2020). Analysis of mechanics of fault reactivation in depleting reservoirs. *International Journal of Rock Mechanics and Mining Sciences, 129*, 104290.

Houwers, M. E., Heijnen, L. J., Becker, A., & Rijkers, R. (2015). A workflow for the estimation of fault zone permeability for geothermal production a general model applied on the Roer Valley Graben in the Netherlands. In *Proceedings World Geothermal Congress 2015*, Melbourne, Australia, 19–25 April 2015.

Ikari, M. J., Saffer, D. M., & Marone, C. (2007). Effect of hydration state on the frictional properties of montmorillonite-based fault gouge. *Journal of Geophysical Research: Solid Earth, 112*(B6), 1–12.

Ishii, E. (2016). The role of bedding in the evolution of meso- and microstructural fabrics in fault zones. *Journal of Structural Geology, 89*, 130–143.

Jolley, S. J., Dijk, H., Lamens, J. H., Fisher, Q. J., Manzocchi, T., Eikmans, H., & Huang, Y. (2007). Faulting and fault sealing in production simulation models: Brent Province, northern North Sea. *Petroleum Geoscience, 13*, 321–340.

Kim, Y. S., & Sanderson, D. J. (2005). The relationship between displacement and length of faults: A review. *Earth-Science Reviews, 68*(2005), 317–334.

Kohli, A. H., & Zoback, M. D. (2013). Frictional properties of shale reservoir rocks. *Journal of Geophysical Research: Solid Earth, 118*, 5109–5125, https://doi.org/10.1002/jgrb.50346

Knott, S., Beach, A., Brockbank, P., Brown, J., McCallum, J., & Welbon, A. (1996). Spatial and mechanical controls on normal fault populations. *Journal of Structural Geology, 18*, 359–372. https://api.semanticscholar.org/CorpusID:129900459

Langhi, L., & Borel, G. D. (2008). Reverse structures in accommodation zone and early compartmentalization of extensional system, Laminaria High (NW shelf, Australia). *Marine and Petroleum Geology, 25*(2008), 791–803.

Lee, H., Jang, Y., Kwon, S., Park, M. H., & Mitra, G. (2018). The role of mechanical stratigraphy in the lateral variations of thrust development along the central Alberta Foothills, Canada. *Geosciences Frontiers, 9*, 1451–1464.

Lindsay, N. G., Murphy, F. C., Walsh, J. J., & Watterson, J. (1993). Outcrop studies of shale smear on fault surfaces. *International Association of Sedimentologists. Special Publications, 15*, 113–123.

Luu, K., Schoenball, M., Oldenburg, C. M., & Rutqvist, J. (2022). Coupled hydromechanical modeling of induced seismicity from CO_2 injection in the Illinois Basin. *Journal of Geophysical Research: Solid Earth, 127*, e2021JB023496. https://doi.org/10.1029/2021JB023496

Manzocchi, T., Childs, C., & Walsh, J. J. (2010). Faults and fault properties in hydrocarbon flow models. *Front Geofluids*, 94–113. https://doi.org/10.1111/j.1468-8123.2010.00283.x

Miocic, J. M., Johnson, G., & Gilfillan, M. V. S. (2020). Stress field orientation controls on fault leakage at a natural CO_2 reservoir. *Solid Earth, 11*, 1361–1374.

Mitchell, T. M., & Faulkner, D. R. (2009). The nature and origin of off-fault damage surrounding strike-slip fault zones with a wide range of dis-placements: A field study from the Atacama fault zone, northern Chile. *Journal of Structural Geology, 31*(8), 802–816. https://doi.org/10.1016/j.jsg.2009.05.002

Morris, A., Ferrill, D. A., & Henderson, D. B. (1996). Slip-tendency analysis and fault reactivation. *Geology, 24*(3), 275–278.

Naruk, S. J., Solum, J. G., Brandenburg, J. P., Origo, P., & Wolf, D. E. (2019). Effective stress constraints on vertical flow in fault zones: Learning from natural CO_2 reservoirs. *AAPG Bulletin, 103*(8), 1979–2008.

Ostanin, I., Anka, Z., & Di Primio, R. (2017). Role of faults in hydrocarbon leakage in the Hammerfest basin, SW Barents Sea: Insights from seismic data and numerical modelling. *Geosciences, 7*(2), 28. https://doi.org/10.3390/geosciences7020028

Pei, Y., Paton, D. A., Knipe, R. J., & Wu, K. (2015). A review of fault sealing behaviour and its evaluation in siliciclastic rocks. *Earth-Science Rev., 150*(October), 121–138. https://doi.org/10.1016/j.earscirev.2015.07.011

Pepin, K. S., Ellsworth, W. L., Sheng, Y., & Zebker, H. A. (2022). Shallow aseismic slip in the Delaware basin determined by Sentinel-1 InSAR. *Journal of Geophysics Research: Solid Earth, 127*(2), e2021JB023157.

Redpath, D., Jackson, C. A.-L., & Bell, R. E. (2022). Mechanical stratigraphy controls normal fault growth and dimensions. Outer Kwanza Basin, offshore Angola. *Tectonics, 41*, e2021TC006823.

Regtien, J., et al. (1995). Interactive reservoir simulation. *Paper SPE 29146 presented at the SPE Reservoir Simulation Symposium.* San Antonio, Texas, USA, February, 12–15.

Roche, V., Camanni, G., Childs, C., Manzocchi, T., Walsh, J., Conneally, J., Saqab, M. M., & Delogkos, E. (2021). Variability in the three-dimensional geometry of segmented normal fault surfaces. *Earth-Science Reviews, 216*, 103523.

Roche, V., Childs, C., Madritsch, H., & Camanni, G. (2020). Layering and structural inheritance controls on fault zone structure in three dimensions: A case study from the northern Molasse Basin, Switzerland. *Journal of the Geological Society, 177–3*, 493–508. https://doi.org/10.1144/jgs2019-052

Ruggieri, R., Scuderi, M. M., Trippetta, F., Tinti, E., Brignoli, M., Mantica, S., et al. (2021). The role of shale content and pore-water saturation on frictional properties of simulated carbonate faults. *Tectonophysics, 807*, 1–12. https://doi.org/10.1016/j.tecto.2021.228811

Schueller, S., Braathen, A., Fossen, H., & Tveranger, J. (2013). Spatial distribution of deformation bands in damage zones of extensional faults in porous sandstones: Statistical analysis of field data. *Journal of Structural Geology, 52*, 148–162.

Shapiro, S. A., Huenges, E., & Borm, G. (1997). Estimating the crust permeability from fluid-injection-induced seismic emission at the KTB site. *Geophysical Journal International, 131*(2), F15–F18. https://doi.org/10.1111/j.1365-246X.1997.tb01215.x

Shapiro, S. A. (2015). *Fluid-induced seismicity*. Cambridge University Press.

Shipton, Z. K., & Cowie, P. A. (2001). Damage zone and slip-surface evolution over μm to km scales in high-porosity Navajo sandstone, Utah. *Journal of Structural Geology, 23*, 1825–1844.

Snippe, J., Kampman, N., Bisdom, K., Tambach, T., March, R., Maier, C., Phillips, T., Forbes Inskip, N., Doster, F., & Busch, A. (2022). Modelling of long-term along-fault flow of CO_2 from a natural reservoir. *International Journal of Greenhouse Gas Control, 118*, 103666. https://doi.org/10.1016/j.ijggc.2022.103666

Takahashi, M., Mizoguchi, K., Kitamura, K., & Masuda, K. (2007). Effects of clay content on the frictional strength and fluid transport property of faults. *Journal of Geophysical Research: Solid Earth, 112*(B8), B08206. https://doi.org/10.1029/2006jb004678

Torabi, A., & Berg, S. S. (2011). Scaling of fault attributes; a review. *Marine Petroleum Geology, 28*(8), 1444–1460.

Tryon, M. D., Henry, P., & Hilton, D. R. (2012). Quantifying submarine fluid seep activity along the North Anatolian Fault Zone in the Sea of Marmara. *Marine Geology, 315–318*, 15–28.

Vadakkepuliyambatta, S., Bünz, S., Mienert, J., & Chand, S. (2013). Distribution of subsurface fluid-flow systems in the SW Barents Sea. *Marine and Petroleum Geology, 43*, 208–221. https://doi.org/10.1016/j.marpetgeo.2013.02.007

Vrolijk, P. J., Urai, J. L., & Kettermann, M. (2016). Clay smear: Review of mechanisms and applications. *Journal of Structural Geology, 86*, 95–152.

Walsh, J., Watterson, J., & Yielding, G. (1991). The importance of small-scale faulting in regional extension. *Letters to Nature, 351*, 391–393.

Ward, N. I. P., Alves, T. M., & Blenkinsop, T. G. (2016). Reservoir leakage along concentric faults in the Southern North Sea: Implications for the deployment of CCS and EOR techniques. *Tectonophysics, 690*, 96–116.

Watson, G. (2020). Faults and fluid flow: An investigation into the production of low permeability fault rock in weakly lithified siliciclastic sequences in New Zealand. Ph.D. thesis, University of Canterbury, Christchurch (New Zealand).

Wenzel, F. (2017). Fluid induced seismicity: Comparison of rate-and-state and critical pressure theory. *Geothermal Energy, 5*(1), 1–16.

Wiprut, D., & Zoback, M. (2002). Fault reactivation, leakage potential, and hydrocarbon column heights in the northern North Sea. *Hydrocarbon Seal Quantification, NPF Special Publication, 11*, 203–219.

Wu, L., Thorsen, R., Ottesen, S, Meneguolo, R., Hartvedt, K., Ringrose, P., & Nazarian, B. (2020). Significance of fault seal in assessing VO2 storage capacity and containment risks—An example from the Horda Platform, northern North Sea). *Petroleum Geoscience, 27*.

Wynants-Morel, N., Cappa, F., De Barros, L., & Ampuero, J. P. (2020). Stress perturbation from aseismic slip drives the seismic front during fluid injection in a permeable fault. *Journal of Geophysics Research: Solid Earth, 125*(7), e2019JB019179.

Yielding, G., Walsh, J., & Watterson, J. (1992). The prediction of small-scale faulting in reservoirs. *First Break, 10*(12).

Yielding, G., Freeman, B., & Needham, D. T. (1997). Quantitative fault seal prediction. *American Association of Petroleum Geologists Bulletin, 81*(6), 897–917. https://doi.org/10.1306/522 B498D-1727-11D7-8645000102C1865D

Zhang, F., An, M., Zhang, L., Fang, Y., & Elsworth, D. (2020). Effect of mineralogy on friction-dilation relationships for simulated faults: Implications for permeability evolution in caprock faults. *Geoscience Frontiers, 11*(2), 439–450. https://doi.org/10.1016/j.gsf.2019.05.014

Zhang, Y., Person, M., Rupp, J., Ellett, K, Celia, M. A., Gable, C. W., Bowen, B., Evans, J., Bandilla, K., Mozley, P., Dewers, T., & Elliot, T. (2013). *Hydrogeologic controls on induced seismicity in crystalline basement rocks due to fluid injection into basal reservoirs* (Vol. 51, No. 4, pp. 525–538). Groundwater–July-August 2013.

Zhao, S., Zheng, M., Liu, S., Li, B., Liu, Y., He, Y., Wang, G., & Qiu, X. (2023). Multiscale fractures and their influence on shale gas deliverability in the Longmaxi formation of the changing block, Southern Sichuan Basin, China. *ACS Omega, 2023*(8), 17653–17666.

Zheng, X., & Espinoz, N. (2022). Stochastic quantification of CO_2 fault sealing capacity in sand-shale sequence. *Marine and Petroleum Geology, 146*, 105961.

Zhou, Q., Birkholzer, J. T., Tsang, C. F., & Rutqvist, J. (2008). A method for quick assessment of CO_2 storage capacity in closed and semi-closed saline formations. *International Journal of Greenhouse Gas Control*, 2(4), 626–639. ISSN 1750-5836, https://doi.org/10.1016/j.ijggc.2008.02.004

Zoback, M. D. (2007). *Reservoir geomechanics* (452 p.). Cambridge University Press. https://doi.org/10.1017/CBO9780511586477

Chapter 2
Influence of Brittleness and Ductility

Abstract In Sect. 2.1, we first focus on the possibility for fault zones to develop a ductile rupture behavior within the brittle upper crust. This is happening to some fault zones affecting clay-rich sedimentary caprock layers. More generally, the brittle-ductile behavior of fault zones has potential strong implications on how fluids can penetrate and circulate within an activated fault. Indeed, a brittle fault will tend to dilate at rupture favoring hydraulic connection between the dilating pores resulting in the creation of fluid flow paths (and leakage). A ductile fault will tend to contract at rupture favoring "closing" of pores connections and preventing flow path creation. For these reasons, in Sect. 2.2 we explore how the brittle-ductile transition can be defined for a given fault zone depending on stress and on the uniaxial compressive strength of fault materials (UCS). In Sect. 2.3, we describe a rare attempt from Piau et al. (J Geophys Res Solid Earth 125, 2020) to adapt a modified Cam-Clay constitutive law to "simulate the effective behavior of fault zones of any maturity, thickness and complexity". Finally, in Sect. 2.4, we apply Piau et al. (J Geophys Res Solid Earth 125, 2020) constitutive law to a scenario of fault reactivation relevant to a CO_2 storage context.

Keywords Brittle-ductile limit · Caprock · Fault mineralogy · Uniaxial compressive strength · Cam-clay physics

2.1 Experimental Insights of Fault Behavior at the Brittle-Ductile Limit

Geological and hydromechanical observations in clay-rich layers and in fault zones show that ductile deformations are possible within the brittle upper crust, i.e., at depths shallower than the limit of the seismogenic zone (Ingram & Urai, 1999; Urai, 1995). The "limit" between brittleness and ductility is defined by the so-called "Mogi line" or "critical state line" that figures the ratio between deviatoric stress and mean effective stress (Fig. 2.1 and see Mogi, 1966 for the experimental approach of ductility; see Roscoe & Burland, 1968 for a theoretical approach of ductility and Wood, 1990 for an exhaustive discussion). Above the line, the rock ruptures by dilatant frictional faulting and rock strength can be described with a linear Mohr–Coulomb friction law or with a nonlinear Hoek–Brown failure envelope for example. Below the line, the rock deforms at the microscale with no clear strain localization on a plane in the form of a fracture or a fault. Eventually a shear band characterized by reorganization of rock minerals/grains and compaction induced by pore collapse will form. A creep law may be used to describe ductile behavior (Aharonov & Scholz, 2018).

The critical state line artificially separates a dilatant and a compacting domain (Fig. 2.2). Observations show that instead of a line there is a progressive transition from brittle-to-ductile properties of rocks. The reason is that in addition to differential and confining stresses, the transition also strongly depends on other conditions such

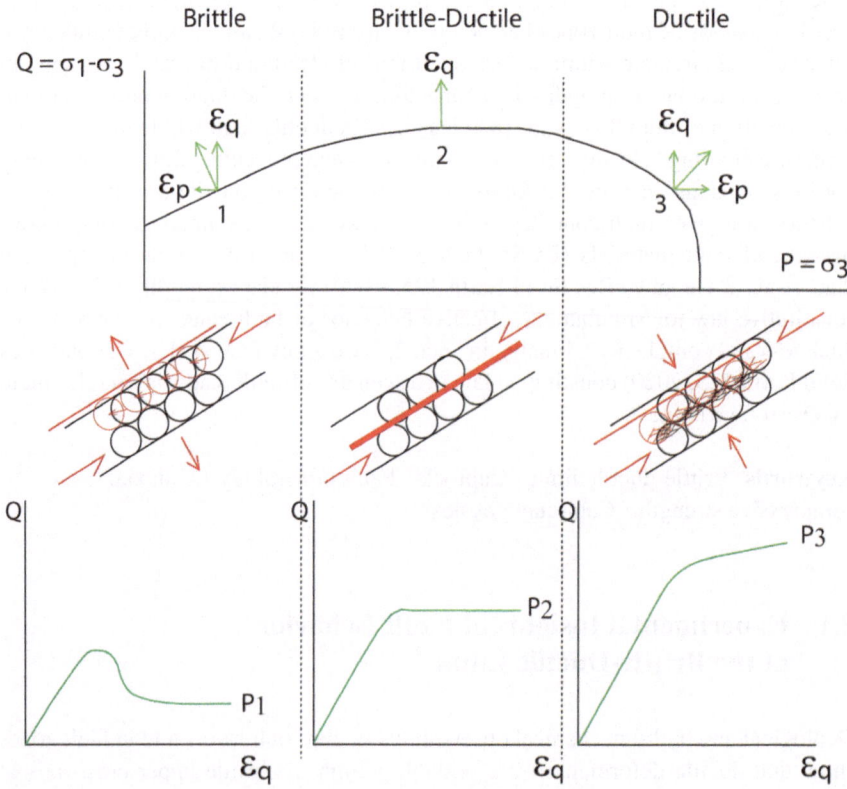

Fig. 2.1 Brittle-to-ductile modes of intact rock fracturing. Upper graph shows the evolution of rock failure from brittle to ductile in a P-Q stress plot. Middle schematic drawings illustrate from left to right, the brittle-dilation of a rock figured as grains, pure isochoric shear at the brittle-ductile transition and compaction damage in ductile mode. Lower graphs show typical triaxial laboratory curves of the same rock sheared at increasing confining stress (P) from left to right, strain softening, pure shear and strain hardening, respectively, from brittle to ductile

as the fluid pressure, the heterogeneity of rock minerals or grains, the heterogeneity of the rock porosity and the cementation between grains that comes from the diagenetic history of the rock. Hirth and Beeler (2015) and Noel et al. (2021) highlight that an increase in pore fluid pressure can progressively bring an initially ductile rock to brittle by dilating the initially compacted pores. In Fig. 2.2, increase in pore pressure will translate the ductile point to the left by decreasing the effective confining stress. Brantut et al. (2008) show that the effect of heterogeneity in the mineral composition of fault rock, each mineral having its onset of plastic deformation occurring under different conditions, favors a semi-brittle behavior that involves the coexistence of both brittle and plastic deformation mechanisms. Among minerals, Jia et al. (2023) show that there is a clear correlation between the brittle-ductile behavior of intact mudstones and the content in clay minerals. They show that above a confining pressure of 5 MPa, an increase of about 10% of the clay mineral content of a mudstone is associated with a 50% decrease in the peak strength, and that high clay content samples display ductile deformation earlier. Plumb (1994) highlights that clay content higher than 35% supports the rock and controls its ductile deformation. Ingram and Urai (1999) and Jia et al. (2023) observe that a cemented shale or mudstone has a higher strength than an uncemented one and is more prone to dilate when deformed at the same stress conditions as an uncemented one.

Direct in situ observations of fault hydromechanical response are rare. Samaroo et al. (2022) compiled a large amount of in situ stress measurements and downhole pore pressures data to show that 72% of the Alberta Basin shale layers were at a brittle-ductile state. They highlighted that almost all the seismicity occurred in brittle shale layers overlying ductile shales over-pressurized by fluid injections. In the same basin, Eyre et al. (2019) documented a distal Mw 4.1 seismic rupture on a shale fault at 3 km depth in the Central Alberta Basin activated by aseismic slip in deeper pressurized brittle-ductile regions of the same fault. In this example, they used rate-and-state-dependent friction laws to describe the spatial decoupling between the aseismic creep and the seismic rupture invoking a change in the fault zone carbonate content favoring seismicity in the upper part. Deeper in the crust, DePaola et al. (2008) and Miller et al. (2004) suggested that a ductile failure initiated by natural CO_2 overpressures in ductile salt may be the trigger of brittle failure and Mw ~ 6 earthquake along faults in overlying carbonates at about 10 km depth in the Northern Apennines. Cheng and Ben-Zion (2019) and Doglioni et al. (2011) show that contrasts in strain rates between the brittle seismogenic crust and the lower ductile crust that take place at about 15 km depths favor seismicity along faults in the upper crust.

At the same time, at the laboratory scale, Meyer et al. (2019) establish that coexistence of localized fault slip with ductile diffuse matrix deformation during the activation of pre-faulted samples of Carrara marble is inversely correlated to the ratio $(\sigma_f - \sigma_y)/(\sigma_{\text{flow}} - \sigma_y)$, where σ_f is the fault strength (related to fault friction), σ_y is the yield stress of the bulk rock (i.e., the minimum stress at which the sample starts deforming plastically, Fig. 2.1) and σ_{flow} is the ductile flow stress. First, these experiments confirm that the brittle-ductile transition starts at the intersection between the

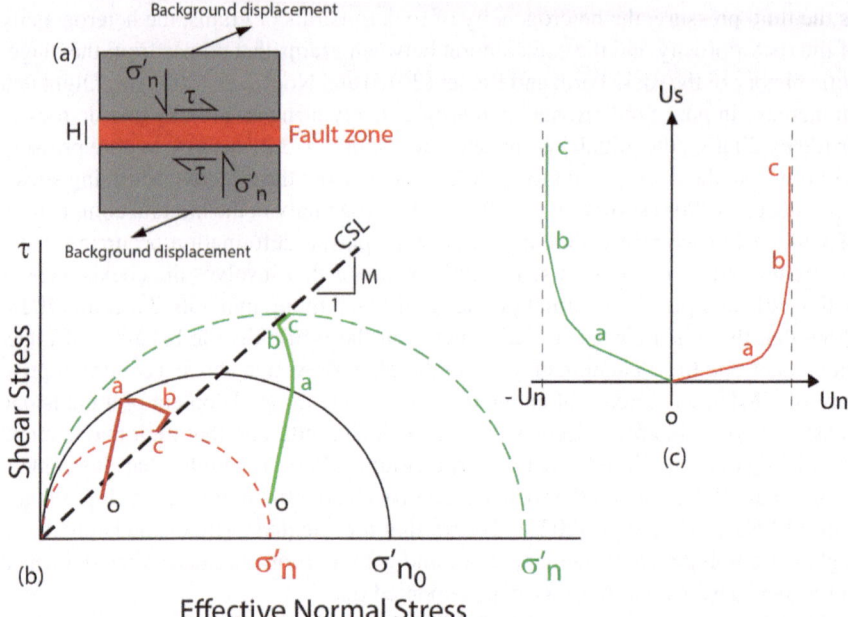

Fig. 2.2 **a** Fault model geometry with a Cam-Clay interface (in Geli et al. 2014 it is a zero thickness interface). **b** Stress path showing two contrasted fault activation mechanisms as developed by Maury et al. (2020). Red path shows a dilatant softening fault rupture. Green path shows a contraction hardening fault rupture. **c** Evolution of the two red and green fault ruptures in a shear (Us)–versus–normal displacement plot (Un). Un < 0 means contraction

matrix yield surface and the critical state line in which slope is related to the fault friction (Fig. 2.2). Second, they show that a significant localized fault slip is still possible in the brittle-ductile transition, even at very high confining stress. In experiments on porous limestone, Noel et al. (2021) show that pore fluid pressure increase immediately turns the fault from contraction to dilation that was initially along localized shear fractures within the fault damage zone. This first generates brittle creep diffuse in the fault zone before localization and slip on a principal macroscopic surface occurs above a given critical dilatancy value. Unfortunately, to our knowledge, there are few observations of fault permeability evolution with the fault brittle-ductile state. Carey and Frash (2017) observed a three order of magnitude permeability decrease with confining stress increasing from 13.5 to 22.2 MPa of sheared Utica shale samples, i.e., while bringing the shales from brittle to brittle-ductile state. Post-mortem samples observations showed that permeability decrease was associated with the transition from a large aperture well-connected fracture network to a distributed, small aperture and poorly connected network. Zhu and Wong (1997) performed exhaustive permeability tests while shearing initially intact 15-to-35% porous sandstones from brittle to ductile/cataclastic flow regime. They systematically observe a drastic permeability decrease associated with grain crushing and pore collapse during the shear band compaction in the brittle-ductile state. Nevertheless, this is less clear/pronounced in low-porosity crystalline rocks.

Analogue water-saturated sandbox experiments explore fault ductility at the meter scale and at extremely low stress compared to the laboratory tests described above. Even if the brittle/ductile deformation mechanisms that are reproduced in these experiments do not occur in the adequate stress-strength conditions, these experiments provide high-resolution observations of the three-dimensional formation of clay smeared fault zones by shearing a low cohesive clay-sand analogue sedimentary pile (Vrolijk et al., 2016). Sand box fault zones architectures are figuring reasonably well the three-dimensional fault zones outcrop observations. These experiments show the impact of brittle-ductile clay smear deformation on across-fault permeability (Kettermann et al., 2017). In details, brittle failure of the clay layer occurs at the onset of faulting, randomly nucleating on some clay layer's heterogeneity. In the patches of brittle fracturing, there is no or little ductile clay smearing, creating "holes" in the clay smear membrane and favoring high across-fault permeability. Experiments suggest that these holes preferentially develop in the footwall cutoff of the clay bed.

2.2 How to Define the Brittle-Ductile Limit?

It is thus important to estimate how far a fault is from the brittle-ductile state limit to anticipate whether the fault may contract aseismically and seal or dilate seismically and leak. There is no universal method to estimate intact rock brittleness which depends on several factors, respectively, the confining stress, lithology and diagenetic history of rocks (Meng et al., 2021 for a review). In addition, to our knowledge, Geli et al. (2014) shows a rare attempt in considering fault rocks brittle-ductile behavior by adapting poro-rigid-plastic constitutive law of Cam-Clay-type (Fig. 2.2; Roscoe et al., 1958). As a starting point, the Mogi or critical state line (Fig. 2.2) provides an empirical brittle-ductile (BD) limit that can be deduced from analyses of stress–strain curves of triaxial laboratory tests performed on various intact rock samples (Walton, 2021). A ductility index d is defined as the slope of the line:

$$d = \frac{\left(\text{differential stress at } BD \text{ limit} = \sigma_1 - \sigma_3^*\right)}{\left(\text{confining stress at } BD \text{ limit} = \sigma_3^*\right)}. \tag{2.1}$$

Walton (2021) considered the intersection between the intact rock yield strength envelope and the critical state line to relate d with Hoek–Brown constant m and the uniaxial compressive strength (UCS).

$$d = \frac{\text{UCS}\sqrt{\frac{\sigma_3^*}{\text{UCS}}m + 1}}{\sigma_3^*} \tag{2.2}$$

Because $\frac{\sigma_3^*}{\text{UCS}}$ does not vary significantly for a given value of d, Walton (2021) suggests a modified ductility parameter $d^* = \frac{d}{\text{UCS}}$ to evaluate the influence of confining stress on intact rock failure mechanisms.

Samaroo et al. (2022) used d and d^* values to devise a method to respectively assess the potential ductile stress state and ductile strength of formations in the Alberta Basin. They identify that most of the induced earthquakes occurred in low confining stress-high differential stress zones (LCS-HDS zones) and in most brittle-ductile layers characterized by $d^* \ll 0.3$ within these zones. They showed that these zones were underlying fluid injection formations highlighting two possible combined driving mechanisms of decreased confining stress through pore fluid pressure increase and increased deviatoric stress through stress transfer.

2.3 Faults Behavior Described with a Cam-Clay Physics

Geli et al. (2014), Maury et al. (2020) and Piau et al. (2020a, 2020b) introduced a Cam-Clay elasto-plastic law to figure the brittle-ductile behavior of a fault interface and reproduce either dilatant/softening or contracting/stiffening of the fault depending on the loading path (Fig. 2.2). The fault rupture envelope is described by:

$$f\left(\sigma_n', \tau, p_c'\right) = \sigma_n'^2 + \sigma_n' p_c' + \frac{3}{M^2}\tau^2, \qquad (2.3)$$

where σ_n' is the effective normal stress, τ the tangential stress, p_c' the yield parameter and M a parameter related to the residual friction angle of the fault. The size of the envelope evolves with p_c' that depends on the displacement normal to the fault surface u_n and on the initial pre-consolidation pressure p_0' (which can be considered as the maximum compressive strength normal to the fault interface in this case):

$$p_c' = p_0' e^{-\alpha \Delta u_n/2}. \qquad (2.4)$$

When $\sigma_{n'}$ and τ are reaching the envelope, the fault normal ($\dot{u}_n$) and slip ($\dot{u}_s$) displacement velocities are given by

$$\dot{u}_n = \dot{K}\left(2\sigma_n' + p_c'\right), \qquad (2.5)$$

$$\dot{u}_s = \dot{K}\frac{6}{M^2}\tau, \qquad (2.6)$$

with K being a scalar called the plastic multiplier (dots above $\dot{u}_n$, $\dot{u}_s$ and $\dot{K}$ are time derivatives).

For a stress path reaching the envelope from below the critical state line, u_n increases (dilatancy) and the envelope size decreases, corresponding to fault weakening. For a stress path reaching the envelope from above the critical state line, u_n

decreases (contractance) and the failure criterion size increases, accounting for fault hardening. At the intersection with the critical state line, the equation is

$$\tau = -\frac{M}{\sqrt{3}}\sigma_{n'}. \tag{2.7}$$

The fault can then slip "infinitely" with constant residual friction angle $\emptyset_c$ ($\tan \emptyset_c = \frac{M}{\sqrt{3}}$), constant thickness $\dot{u}_n = 0$ and constant failure envelope size.

Using a Cam-Clay fault interface could thus allow reproducing brittle-ductile fault deformations with a relatively limited number of fault input properties, respectively $p_{0'}$, M and α. In Chap. 4, we discuss the possibility offered by this approach to define a stable-to-unstable fault slip evolution, and thus to describe different types of mechanisms such as slow and seismic slip (the instability being defined in the formulation of the plastic multiplier K). The question is then to estimate relevant values for $p_{0'}$, M and α. There is a reasonable dataset of residual frictional angle ($\emptyset_c$) that can help determine M from the $\left(\tan \emptyset_c = \frac{M}{\sqrt{3}}\right)$ relationship.

The pre-consolidation pressure $p_{c'}$ refers to the maximum compressive strength normal to the fault interface. In the case of an active fault $f(\sigma_{n'}, \tau, p_{c'}) = 0$, $p_{c'}$ can be defined from Eq. 2.6 (given the Cam-Clay criterion defined by Piau et al. 2020a, 2020b):

$$p_{c'} = -\left(\sigma_{n'} + \frac{3}{M^2}\frac{\tau^2}{\sigma_{n'}}\right). \tag{2.8}$$

With $\sigma_{n'}$ and τ respectively are the active normal and shear stresses applied on the fault interface.

In the case of an inactive fault, one possibility is to produce a local reactivation of the fault by injecting high-pressure fluid. This is possible through "hydroshearing" tests conducted in boreholes cross-cutting fault zones. Example of such hydroshearing experiments are given in Sect. 3.5. Equation (2.8) would then be:

$$p_{c'} = -\left((\sigma_{n'} + \Delta P_f) + \frac{3}{M^2}\frac{\tau^2}{(\sigma_{n'} + \Delta P_f)}\right), \tag{2.9}$$

where ΔP_f is the fluid overpressure to activate the fault.

The initial pre-consolidation pressure $p_{0'}$ refers to the maximum pressure ever experienced by the fault zone during its geological history. In the case of a sedimentary layer, it can be related to the maximum effective burial depth of the layer $\sigma'_{v\,max}$ if the sediment consolidation is mainly due to vertical mechanical compaction. $p_{0'}$ is also related to the over-consolidation ratio (OCR) through the equation OCR $= \frac{p_{0'}}{p} = \frac{\sigma_{v\,max'}}{\sigma_{v'}}$, where $p = \sigma_{v'}$ is the effective vertical stress at the actual layer depth. In shales and clays, the OCR has been empirically correlated to the normalized undrained shear strength of the intact rock by Nygård et al. (2006).

$$\frac{(\sigma_1 - \sigma_3)}{2\sigma_v'} = a(\text{OCR})^b, \tag{2.10}$$

with $a = 0.39$ and $b = 0.89$.

In addition, several authors consider that for OCR > 2.5, mudrocks and clay rocks are brittle and prone to develop fluid leakage flow paths under shear fracturing (Nygård et al., 2006; Ingram & Urai, 1999). This means that for clay rocks, the brittle-ductile limit would be corresponding to a $p_0' \leq 2.5\sigma_v'$.

With some rearranging, we can get the following empirical equation for p_0' as a function of the deviatoric and the confining stresses.

$$p_0' = \sigma_v' e^{\frac{\ln\left(\frac{\sigma_1 - \sigma_3}{2a\sigma_v'}\right)}{b}} \tag{2.11}$$

Nevertheless, getting values of p_c', p_0' and α for fault zone material may require the inversion of fault displacements or strains measured in situ (to our knowledge there currently are very few of such available datasets). Moreover, determination of the shape of the Cam-Clay criterion $f(\sigma_n', \tau, p_c')$ and calibration of the p_c' the yield parameter evolution with fault deformation will require additional research (beyond the scope of this review). In Sects. 2.4 and 4.4, we will limit ourselves to conducting a sensitivity study to Cam-Clay parameters for fault environments relevant to various CO_2 storage contexts.

2.4 Sensitivity Study of a Fault Rupture to Cam-Clay Parameters

The Cam-Clay interface has been developed by Maury et al. (2020) to describe the complex rupture propagation on crustal faults at depths larger than 4 km and relatively large 10 cm/yr tectonically active far-field compressive regimes. In this section, we explore how this type of constitutive law can inform on rupture propagation on a relatively small, 3–5 km long, 10 m thick and 70° dipping fault zone intersecting a CO_2 storage layer at a depth of 2, 3 km (Fig. 2.3 and Table 2.1). As in Maury et al. (2020), the fault zone is simplified as a constant thickness layer that potentially includes both a core and a damage zone. We consider an average fault zone porosity of 0.12, and the following Cam-Clay parameters $(M, P_0', \alpha) = (0.7, 45, 21)$ that correspond to a fault residual friction angle of ~ 22°, $p_0' \sim 2.2\sigma_v'$ and a moderate dilatant fault zone. The initial shear and effective normal state of stress on the fault is 7.7 and 16 MPa, with an initial brine pore pressure of 20 MPa. This corresponds to a point located on the fault at 2.3 km depth under stress components Sxx = 34 MPa, Syy = 43 MPa and Szz = 53 MPa (the fault strike being perpendicular to Sxx, Fig. 2.3a). The fault is initially stable. We consider a brine compressibility of 0.00046 MPa^{-1}.

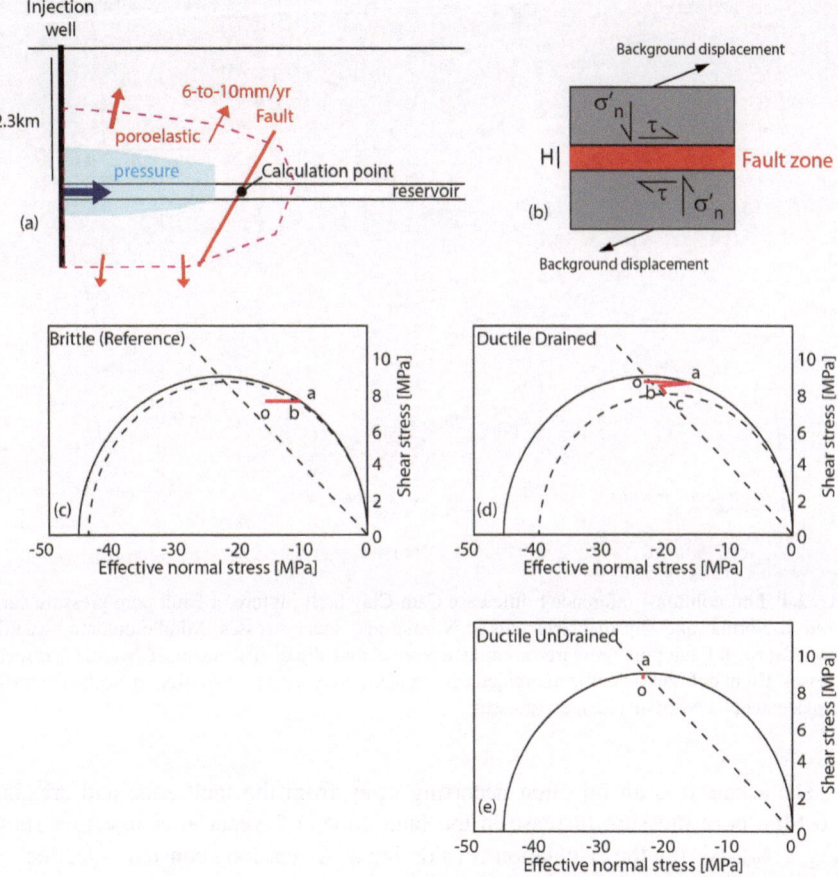

Fig. 2.3 **a** Schematic vertical geological cross-section of the case study. **b** Cam-Clay criterion used in this study. **c** Reference brittle case with the red line figuring the stress path showing the elastic loading phase (oa) that is associated to pressure build up in the fault and the plastic fault rupture phases (ab). **d** Ductile drained case with the stress path showing the elastic loading oa starting at o that is below the critical state line in the ductile domain, pressure driven rupture ab and bc. **e** Ductile undrained case with stress path showing elastic loading oa and rupture at a

Table 2.1 Cam-Clay fault reference model properties

Intact rock		Fault zone		Initial stress	
ρ (kg/m^3)	2500	Thickness (m)	10	τ (MPa)	7.7
E (MPa)	5800	Porosity m	0.12	σ'_n (MPa)	16
ν	0.29	M	0.7	Pore pressure	
		P'_0 (MPa)	45	Pf_0 (MPa)	20
		α	21		
		Fluid compressibility		Background displacement rate	
		C_{fl} (MPa^{-1})	0.00046	0.006 m/yr	

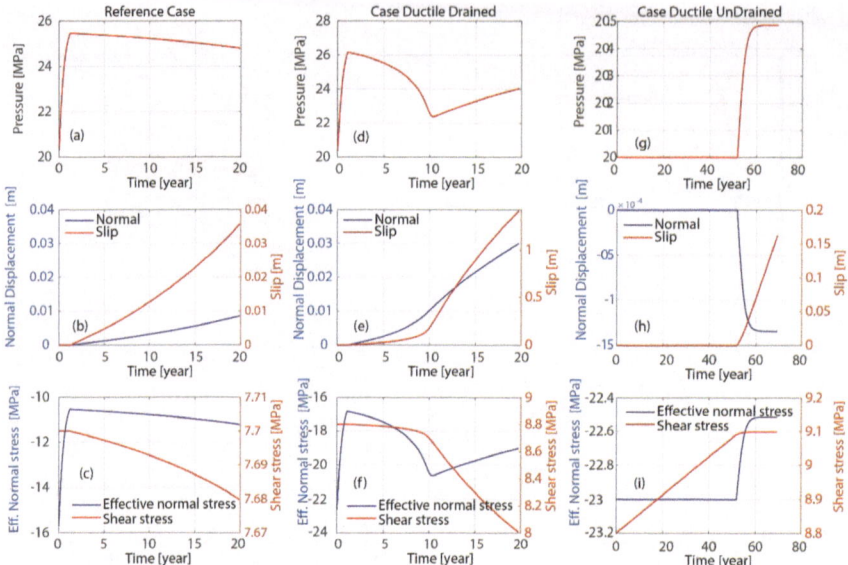

Fig. 2.4 Left column—reference brittle case Cam-Clay fault rupture. **a** Fault pore pressure variation. **b** Normal and slip displacements. **c** Normal and shear stresses. Middle column—ductile drained case. **d** Fault pore pressure variation. **e** Normal and slip displacements. **f** Normal and shear stresses. Right column—ductile undrained case. **g** Fault pore pressure variation. **h** Normal and slip displacements. **i** Normal and shear stresses

The scenario is an injection occurring away from the fault zone and creating a 6 MPa pore pressure increase in the fault zone, 1.2 years after injection starts (Fig. 2.4a). We run the calculation considering a 20 year long constant injection.

In our calculations, we use the poro-plastic hardening/weakening interface constitutive law version 1.1 developed by Piau et al. (2020a, 2020b). The fault is loaded by the pressure increase in the reservoir that results in a decrease of the fault initial effective normal stress that lasts 1.2 years ((oa) stress path in Figs. 2.3a and 2.4a). This simple analytical model does not calculate the fault elastic response. The fault pore pressure variation is described by Eq. 2.12:

$$\Delta P_f = -S \Delta u_n + R(P_{\text{injection}} - P_{\text{initial}}). \qquad (2.12)$$

First term of the equation depends on parameter S that is coupling the change in the fault pore pressure to the fault dilation Δu_n. Second term of the equation figures the fluid injection as an external pressure source leaking into the fault zone. R is a coefficient relating fault pore pressure variation to the external fluid source in drained conditions (units in year). Here we use S formulation from Piau et al. (2020a, 2020b) where the roles of fault thickness, porosity and fluid compressibility are considered together in Eq. 2.13:

$$S = \frac{1}{H \emptyset C_{fl}}, \tag{2.13}$$

where H is the fault thickness, $\emptyset$ is the fault porosity and C_{fl} is the fluid compressibility.

The chosen initial state of stress places the fault in the brittle domain (point O in Fig. 2.3c). Increase in fault pore pressure due to injection induces a decrease in the effective normal stress until reaching the failure envelop at point a (Fig. 2.3c) and 1.2 years after injection start (Fig. 2.4a). After rupture, a slow dilatant softening period, corresponding to a shrinking of the failure envelop, spans until the end of the calculation at point b (Fig. 2.3c). Until 20 years, this dilatant softening fault rupture translates into slow slip with an associated normal opening (Fig. 2.4b). The slow fault normal opening explains the slow pore pressure drop (Fig. 2.4a) and the associated increase in effective normal stress ((ab) stress path in Fig. 2.3c and time variations in Fig. 2.4c). The total slip and normal opening after 20 years of injection are 38 × 10^{-2} m and 9 × 10^{-3} m, respectively associated to a pore pressure drop of 1 MPa (Fig. 2.4b).

We then considered the same fault properties but in a slightly higher stress context characterized by a shear stress of 8.8 MPa and an effective normal stress of 23 MPa. This case corresponds to a deeper point of the fault that is in the ductile domain compared to the previous one (Fig. 2.3d–e). In addition, we consider two ductile fault rupture cases, drained and undrained (Figs. 2.3d–e and 2.4d–i). The drained case (Figs. 2.3d and 2.4d–f) figures a hydraulic pressure connection between the point and the injection while the undrained case (Figs. 2.3e and 2.4g–i) figures a point of the fault that is not hydraulically connected to the injection.

In the drained ductile case, the fault pore pressure increase reduces the effective normal stress and moves the fault from the ductile to the brittle domain (from point o to point a in Fig. 2.3d). At failure the fault response is the same during the first 10 years as in the dilatant-brittle case (points a to b in Fig. 2.3d). Increase in normal displacement associated to slip generates dilation and decrease in pore pressure (Fig. 2.4e–d). Softening corresponds to a slight decrease in shear stress while the decrease in pore pressure induces an increase in the effective normal stress (Fig. 2.4f). At 10 years of injection, stress path on the fault is getting close to the critical stress line (point b in Fig. 2.3d) and follows the line until 20 years (point c in Fig. 2.3d). The CSL is reached in the ductile drained case and not reached in the brittle drained case because the stress variation to reach the CSL is smaller in the ductile than in the brittle case (compare Fig. 2.3c, d). Along the critical stress line (from c to d in Fig. 2.3d), the fault response becomes more frictional (close to a Mohr–Coulomb failure case). There is a significant increase in slip that significantly exceeds the increase in normal displacement (Fig. 2.4e). At this time, fault mainly slips with relatively less opening compared to before reaching the CSL. This explains the increase in the pore pressure and the decrease of the effective normal stress (Fig. 2.4f). In addition, shear stress is decreasing faster. At 20 years, the calculated fault slip of 1.4 m is much larger than the 38 × 10^{-2} m slip in the previous brittle case (Fig. 2.4b–e).

In the undrained ductile case, the fault is slowly brought to failure by a far-field displacement rate of 0.006 m/year (oa in Fig. 2.3e). To justify this displacement rate, we consider that as soon as injection starts, a strong deformation develops in the pressure build up area causing the storage reservoir poroelastic expansion (Fig. 2.3a). We hypothesize that this expansion triggers a small 0.006 m/year displacement almost parallel to the fault and a small 0.0001 m/year compressive displacement perpendicular to the fault. We hypothesize that such displacement occurs way beyond the pressurized volume in areas of the fault that are not hydraulically connected to the injection pressure source (Fig. 2.3a). Given such conditions, fault failure occurs at 56 years after calculation start (Fig. 2.4g–i). We then calculate fault failure for about 15 years (calculation is stopped at 71 years). During this period, the fault tends to compact (negative normal displacement) while slipping (Fig. 2.4h). The small 1.5×10^{-4} m compaction generates a 0.5 MPa fault pore pressure increase (Fig. 2.4g) and an associated decrease in the effective normal stress (Fig. 2.4i).

We then conducted a sensitivity study of fault movements to model's parameters in order to identify how they influence the fault slip and normal displacement evolution versus time (Figs. 2.5, 2.6 and 2.7). In this sensitivity study, we consider an initial brittle state of stress on the fault (Fig. 2.3c) that is the same for all cases, while we focus on varying the following parameters that can potentially influence the weakening of the fault and change its slip and normal displacement:

- The Cam-Clay parameters p_0', M and α that respectively define the size and shape of the initial failure envelop and the evolution of the initial pre-consolidation pressure p_0' with the variation of the fault normal displacement (Fig. 2.5).

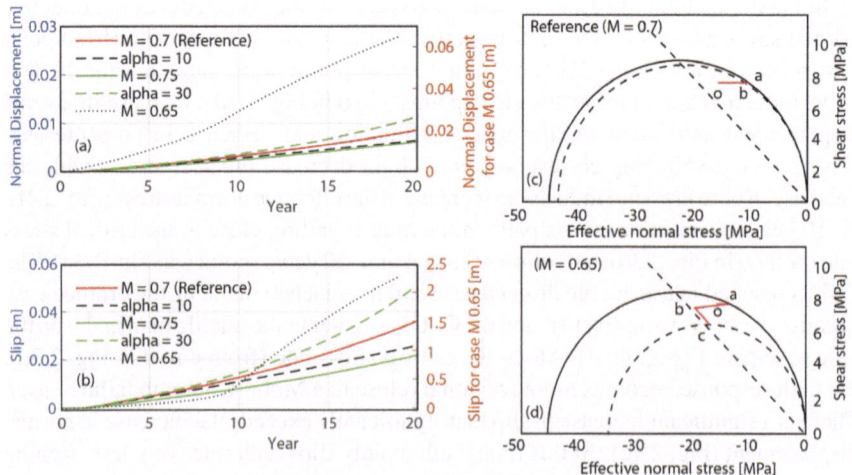

Fig. 2.5 Effects of Cam-Clay constitutive law parameters on fault slip and normal displacement. **a** Normal displacement. **b** Slip. **c** Stress path for the reference $M = 0.7$ and alpha = 21 case. **d** Stress path for the case with $M = 0.65$

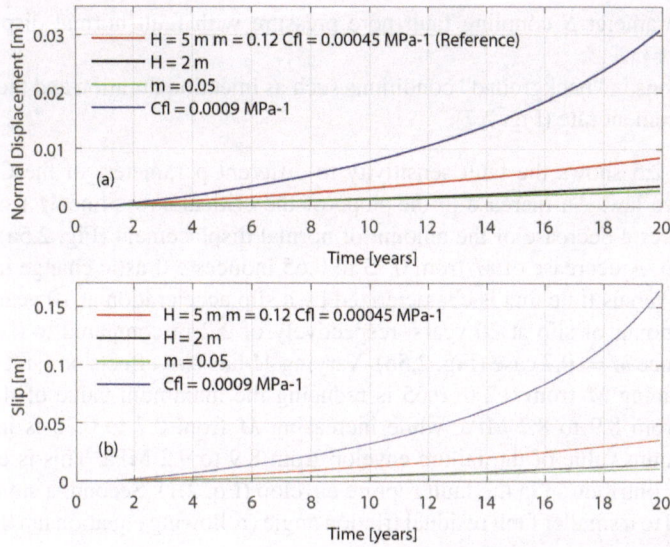

Fig. 2.6 Effects of fault thickness, porosity and pore fluid compressibility on fault zone rupture. **a** Normal displacement. **b** Slip

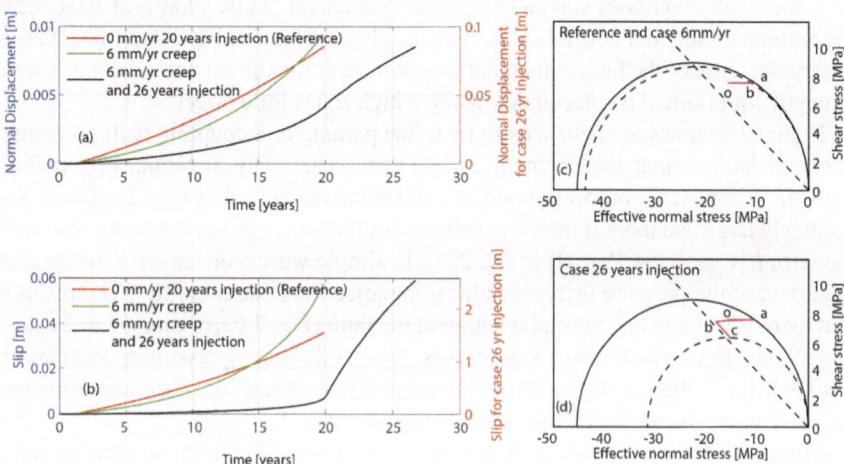

Fig. 2.7 Effects of the background displacement rate and of the injection duration. **a** Normal displacement. **b** Slip. **c** Reference case stress path (corresponds to red curves in a–b). **d** Effect of the background displacement rate combined with a longer injection on the stress path (corresponds to black curves in a–b)

- The parameter S coupling fault pore pressure with fault normal displacement (Fig. 2.6).
- Variations in "background" conditions such as injection duration and the far-field displacement rate (Fig. 2.7).

Figure 2.5 shows the fault sensitivity to different parameters of the Cam-Clay constitutive law. An increase in the slope of the critical stress line M from 0.7 to 0.75 induces a decrease of the amount of normal displacement (Fig. 2.5a) and slip (Fig. 2.5b). A decrease of M from 0.75 to 0.65 induces a drastic change in the slip evolution versus time that is characterized by a slip acceleration at 10 years and by a large amount of slip at 20 years, respectively of 2.2 m compared to 0.038 m in the reference $M = 0.7$ case (Fig. 2.5b). Varying M has two effects on fault strength. First, reducing M from 0.7 to 0.65 is reducing the maximum value of the failure envelop from 8.9 to 8.2 MPa, while increasing M from 0.7 to 0.75 is increasing the maximum value of the failure envelop from 8.9 to 9.2 MPa. This is explained by the M contribution in the fault rupture envelop (Eq. 2.1). Second, a smaller M is associated to a smaller fault residual friction angle (following equation $\tan \emptyset_c = \frac{M}{\sqrt{3}}$). Overall, a low M value reflects a low strength fault zone with a larger weakening at rupture. A variation of the α coefficient in Eq. 2.2 changes the evolution of the fault consolidation pressure (p_c') with normal displacement variation. A high α of 30 significantly increases slip and normal displacement, while a low α of 10 reduces the amount of slip and normal displacement. This is because, Eq. 2.4, a low α favors a large decrease of the fault consolidation pressure or maximum normal compressive strength with normal displacement, while a high α has little effect.

Figure 2.6 shows the fault sensitivity to the parameter S coupling fault pore pressure with fault normal displacement (Eq. 2.11). If there is a lower volume of pore fluid in the fault zone (case of low porosity $m = 0.05$ and case of a thin $H = 2$ m fault), S is relatively large and there is more contribution of the term $-S \Delta u_n$ in comparison with the term $R(P_{injection} - P_{initial})$ in Eq. 2.10. In simple words, the larger S the smaller the effect of the pressure increase induced by injection (less coupling). The result is that there is less slip and normal opening of the fault (Fig. 2.6a, b, cases $H = 2$ m and $m = 0.05$). In contrast, when S decreases under a fluid compressibility increase to 0.0009 MPa^{-1}, the injection pressure increase effect on fault displacement becomes larger. The results are larger slip and normal opening.

Figure 2.7 shows the effects of different "background" conditions such as injection duration and far-field displacement rate. An increase of 6 mm/year of the rate of the displacement tangential to the fault increases the slip and normal displacement. Indeed, a part of this displacement is being converted into slip and added to the slip induced by the increase in the fault pore pressure. Not shown in Fig. 2.7, background displacements perpendicular to the fault plane have less impact on rupture. An increase of 6 mm/year of the tangential displacement rate can for example be caused by rock poroelastic stressing around a CO_2 storage injection project. If the injection lasts longer than 20 years, the combination of high pore pressure and background displacement rate may lead the fault softening to the critical stress line as shown in Fig. 2.7 (black curve). As described before, when the critical stress line is reached there is a significant increase in slip and normal displacement rates.

2.5 Conclusions

The Cam-Clay plastic framework gives a large end-cap plasticity model that can integrate all types of fault behaviors. In the dilatant-brittle domain, faults rupture affecting low clay content rocks may be described. In the ductile domain, large clay content fault rocks and high confining stress conditions that may respectively occur in caprocks and/or in deep rocks can be considered. This cannot be done with classical Coulomb criterion that has no end-cap, thus cannot describe ductile failure. Moreover, the complex fault rock behavior at the brittle-ductile transition can also be considered. This may be of great interest to describe how a caprock in initial strengthening ductile state can move to a softening brittle and dilatant state under an increase in pore pressure produced by storage in a nearby reservoir. Although the Cam-Clay criterion requires some input parameters possibly as tricky to estimate as the Coulomb criterion, it offers the possibility to consider a much broader range of fault leakage and rupture scenarios. The sensitivity study of a "stable" fault activation caused by a CO_2 storage injection showed results that are consistent with observations at basin scales. Moreover, it highlights the strong influence of some parameters such as the change in fluid content, the residual strength and initial fault consolidation state on fault hydromechanical response. It also shows that slow slip fault evolution lasting over decades can be triggered by an increase of the background strain rate and the connection or no-connection of the fault to the injection source. In Fig. 2.8, we show how some of these effects combined together can significantly accelerate fault slip.

In comparison, Mohr–Coulomb does not inform on the evolution of the slip velocity and may overestimate the dilation of the fault since it is often fixed by an arbitrary dilation angle initial value which is intrinsic to a non-associated Coulomb flow law. The Cam-Clay gives more flexibility for the dilation to evolve from failure initiation of the fault figured as a bulk zone until the fault reaches the critical stress line where dilation may only be constrained on a localized frictional interface.

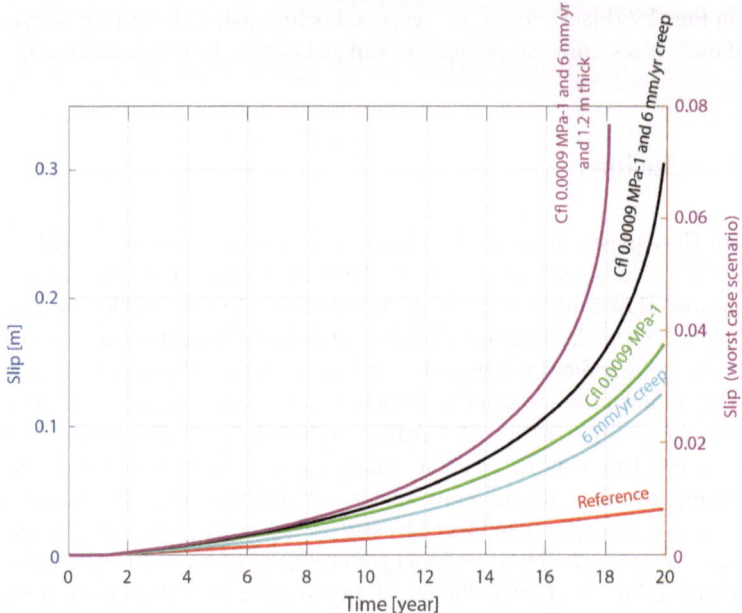

Fig. 2.8 How combined fault, fluid properties and background creep rates can accelerate fault slip in a context relevant to a CO_2 storage injection of 20 years (except the worst case scenario, all the other fault cases correspond to a 5 m thick fault zone)

References

Aharonov, E., & Scholz, C. H. (2018). The Brittle-Ductile transition predicted by a physics-based friction law. *Journal of Geophysical Research: Solid Earth, 124,* 2721–2737.

Brantut, N., Schubnel, A., Rouzaud, J. N., Brunet, F., & Shimamoto, T. (2008). High-velocity frictional properties of a clay-bearing fault gouge and implications for earthquake mechanics. *Journal of Geophysics Research: Solid Earth, 2008*(113), 1–18.

Carey, J. W., & Frash, L. P. (2017). Brittle-ductile behavior and caprock integrity. *Energy Procedia, 114,* 3132–3139.

Cheng, Y., & Ben-Zion, Y. (2019). Transient Brittle-Ductile transition depth induced by moderate-large earthquakes in Southern and Baja California. *Geophysical Research Letters, 46,* 11109–11117.

De Paola, N., Collettini, C., Faulkner, D. R., & Trippetta, F. (2008). Fault zone architecture and deformation processes within evaporitic rocks in the upper crust. *Tectonics, 2008*(27), 1–21.

Doglioni, C., Barba, S., Carminati, E., & Riguzzi, F. (2011). Role of brittle-ductile transition on fault activation. *Physics of the Erath and Planetary Interiors, 184,* 160–171.

Eyre, T. S., Eaton, D. W., Garagash, D. I., Zecevic, M., Venieri, M., Weir, R., & Lawton, D. C. (2019). The role of aseismic slip in hydraulic fracturing-induced seismicity. *Science Advances, 5.*

Geli, L., Piau, J. M., Dziak, R., Maury, V., Fitzenz, D., Coutellier, Q., & Henry, P. (2014). Seismic precursors linked to super-critical fluids at oceanic transform faults. *Nature Geosciences, DOE.* https://doi.org/10.1038/NGEO2244

Hirth, G., & Beeler, N. M. (2015). The role of fluid pressure on frictional behavior at the base of the seismogenic zone. *Geology, 43,* 223–226.

Ingram, G. M., & Urai, J. L. (1999). Top-seal leakage through faults and fractures: the role of mudrock properties. In A. C. Aplin, A. J. Fleet & J. H. S. Macquaker (Eds.), *Muds and mudstones: Physical and fluid flow properties* (pp. 125–135). Geological Society, London, Special Publications, 158.

Jia, R., Fu, X., Jin, Y., Wu, T., Wang, S., & Cheng, H. (2023). Mechanical properties of mudstone caprock and influencing factors: Implications for evaluation of caprock integrity. *Frontiers in Earth Science, 11*, 1229851.

Kettermann, M., Urai, J. L., & Vrolijk, P. J. (2017). Evolution of structure and permeability of normal faults with clay smear: Insights from water-saturated sandbox models and numerical simulations. *Journal of Geophysics Research: Solid Earth, 122*, 1697–1725.

Maury, V., Piau, J.-M., & Fitzenz, D. (2020). Conditions for triggering seismic ruptures and/or slow slip events in the framework of a poro-plastic fault zone model. *Journal of Geophysical Research: Solid Earth, 125*.

Meng, F., Wong, L. N. Y., & Zhou, H. (2021). Rock Brittleness indices and their applications to different fields of rock engineering: A review. *Journal of Rock Mechanics and Geotechnical Engineering, 13*, 221–247.

Meyer, G. G., Brantut, N., Mitchell, T. M., & Meredith, P. G. (2019). Fault reactivation and strain partitioning across the brittle-ductile transition. *Geology, 47*, 1127–1130.

Miller, S. A., Colletini, C., Chiaraluce, L., Cocco, M., Barchi, M., & Kaus, B. J. P. (2004). Aftershocks driven by a high-pressure CO_2 source at depth. *Nature, 427*.

Mogi, K. (1966). Pressure dependence of rock strength and transition from brittle fracture to ductile flow. *Bulletin Earthquake Research Institute, University of Tokyo, 1966*(44), 215–232.

Noel, C., Passelegue, F. X., & Violay, M. (2021). Brittle faulting of ductile rock induced by pore fluid pressure build-up. *Journal of Geophysical Research: Solid Earth, 126*, e2020JB021331.

Nygård, R., Gutierrez, M.,Bratli, R. K., & Høeg, K. (2006). Brittle–ductile transition, shear failure and leakage in shales and mudrocks. *Marine and Petroleum Geology, 23* (2006), 201–212.

Piau, J.-M., Maury, V., Fitzenz, D. (2020). READ ME Supplementary materials proposed for CP1 and CP2.docx. *Modelling of the mechanical behaviour of active geological faults using a poro-plastic hardening/weakening interface constitutive law*, data.univ-gustave-eiffel, V1.

Piau, J.M., Maury, V., & Firenz, D. (2020). Interface plastic constitutive law with end-cap and structural model applied to geological faults behavior. *Journal of Geophysical Research: Solid Earth, 125*.

Plumb, R. (1994). Influence of composition and texture on the failure properties of clastic rocks. In SPE rock mechanics in Petroleum engineering (Delft, Netherlands SPE), 13–20.

Roscoe, K. H., Schofield, A. N., & Wroth, C. P. (1958). On the Yielding of Soils. *Géotechnique, 8*, 22–53. https://doi.org/10.1680/geot.1958.8.1.22

Roscoe, K. H., & Burland, J. B. (1968). In J. Heyman & F. A. Leckie (Eds.), *On the generalized stress-strain behaviour of wet clay*. Cambridge at the University Press.

Samaroo, M., Chalaturnyk, R., Dusseault, M., Chow, J. F., & Custer, H. (2022). Assessment of the Brittle-Ductile state of major injection and confining formations in Alberta Basin. *Energies, 15*, 6877.

Urai, J. L. (1995). Brittle and ductile deformation of mudrocks. EOS Transactions, American Geophysical Union, November 7 1995, F565.

Vrojlik, P. J., Urai, J. L., & Kettermann, M. (2016). Clay smear: Review of mechanisms and applications. *Journal of Structural Geology, 86*, 95–152.

Walton, G. (2021). A new perspective on the Brittle-Ductile transition of rocks. *Rock Mechanics and Rock Engineering, 54*, 5003–6006.

Wood, D. (1990). *Soil behaviour and critical state soil mechanics*. Cambridge University Press.

Zhu, W., & Wong, T. F. (1997). The transition from brittle faulting to cataclastic flow: Permeability evolution. *Journal of Geophysical Research, Vil., 102*(B2), 3027–3041.

Chapter 3
Factors Influencing Fault Permeability Evolution During and After Failure (Sealing)

Abstract In this chapter, we explore the different factors influencing fault permeability development during fracturing and sealing periods. In underground CO_2 storage, the large-scale fluid pressure changes expected from industrial injections need to be kept lower, by regulation, than the intact rock fracturing pressure to avoid geomechanical damage in the reservoir and hydrofracturing in the confining caprock units. Nevertheless, these injection pressures may be high enough to activate dilatant shearing on natural pre-existing faults intersecting the reservoir-caprock system (Guglielmi et al. in Int J Greenhouse Gas Control 111, 2021). This could result in the creation of permeable flow pathways along the faults which could allow for CO_2 to penetrate and migrate through the overlying sealing formations. Unless caprock clay-rich fault self-sealing occurs, the CO_2 could escape from the reservoir, and eventually end up in overlying fresh-water aquifers or leak to the surface, thus creating a risk to the environment and defeating the sequestration purpose. In Sects. 3.1, 3.2, 3.3 and 3.4, we review different factors influencing single-fractures permeability evolution under shear, reporting observations at laboratory scale. In Sect. 3.5, we explore how permeability of a fault zone can vary under activation at field scale. This chapter summarizes the key results from several field scale fault activation experiments. In these experiments, fluid pressure was increased through fluid injections conducted directly into natural fault zones affecting a wide variety of rock geology. In Sect. 3.6, we suggest a conceptual model of a fault zone permeability evolution with fault plastic rupture, and we report on some permeability laws that could help to consider such evolution in advanced numerical models.

Keywords Fault permeability · Clay content · Stress/strength · Slip/dilation · Laboratory-versus-field scale experiments · Sealing · Constitutive permeability laws

3.1 Effect of Fault Material Characteristics: Clay Mineral Content and Fabric

It is commonly admitted that the amount of clay content drives the initial permeability (i.e., permeability before fault mechanical activation) of a fault zone (see for example Yielding et al., 1997) and the permeability change at fault activation. In sandstones or carbonates, the fault clay content varies between the fault core and the fracture damage zone (Fig. 3.1). In shale faults, the entire fault zone displays a high clay content. Below about 20% of clay content, fault permeability varies over a broad range of values from 10^{-16} to 10^{-11} m^2. In these low clay content fault materials, the permeability at brittle fault activation and shearing is complex. Farrell et al. (2021) observe that shearing destroys the authigenic clay minerals coating the sandstone pores, breaks the capillary barriers and results in 2–3 orders permeability increase. To the contrary, Zhu and Wong (1997) observe that dilatant shearing of a > 15% porosity sandstone results in a permeability decrease, that they explain by an increase in the tortuosity caused by intense micro fracturing. The same authors observe a permeability increase during dilatant shearing of low-porosity crystalline rocks. Above 20% of clay content, fault permeability is very low, respectively of 10^{-17} to 10^{-20} m^2. Permeability may show a slight decrease trend with the increase in clay content but the 20% clay content looks more like a threshold between high- and low-permeable fault zones. In the extreme case, faults within shale layers or caprocks tend to have a permeability close to that of intact clay rock (Jeanne et al., 2018). In addition, swelling of clay minerals can drastically reduce and even completely seal a fracture (Fang et al., 2017).

3.2 Effect of Fault Asperities Geometry and Strength

Laboratory observations indicated that the size, shape and strength of fracture asperities influence the distribution and evolution of permeability and flow paths (Ye & Ghassemi, 2018, 2019; Ji et al., 2023). For instance, Barton et al. (1985) and Chen et al. (2000) observe 1-to-3 orders of permeability increase associated to 6-to-60 mm slip-induced dilation on different types of rough fractures and from low 0.0025 up to over 40 MPa effective normal stress (note that at laboratory scale the term "fracture" is often used even if the experiments described in this chapter focus on fracture shearing. The term "fracture" at laboratory scale comes from the initial setting of these experiments in which an intact rock sample is cracked with no movement, and then set under shearing conditions). They relate dilation to the fracture roughness amplitude and shape, and to the effective normal stress applied on the fracture surface. In addition, Chen et al. (2000) highlight the strong tortuosity effect played by fractures asperities. It also appears that the permeability change with shear displacement follows a negative power law, meaning that there is an upper limit to permeability increase under shear related to the amplitude of fracture

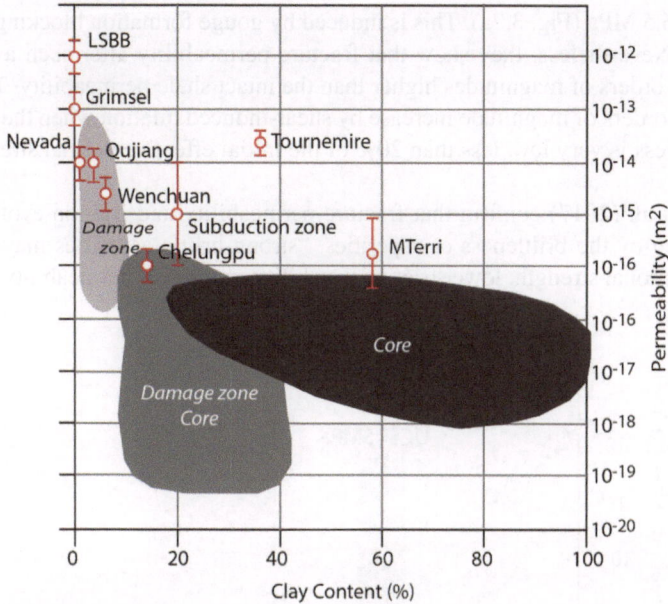

Fig. 3.1 Relationship between permeability and clay content in different types of faults affecting sandstones (modified from Farrell et al., 2021). In red, we added values from some of the sites in carbonates and shales, mentioned in the text

asperities. Nevertheless, laboratory tests conducted at higher confining pressures and higher total shear displacements than Barton et al. (1985) pioneering tests, all tend to show permeability decreases under shear with complex evolutions of fractures. For example, after producing a fault slip of 18mm on saw-cut laboratory samples of low porosity Novaculite and Diorite under effective normal stresses of 5–20 MPa, Faoro et al. (2009) observe 3 orders of magnitude decrease of macro-fractures permeability associated to gouge formation by asperity breaking. When a gouge "layer" is formed, Zhang et al. (1999) observe some complexity characterized by a general tendency of a one additional order of permeability decrease with shearing of gouge materials but with some possible permeability increases depending on gouge content in phyllosilicate minerals. Thus, the destruction of fracture roughness or asperities under high effective normal stress generates gouge and permeability reduction that balances the slip-induced dilatant effect related to sliding on intact asperities (Li et al., 2023). There is a complex relationship between asperities amplitude and strength and fracture permeability change under shear as observed at the laboratory scale. A low asperity tensile and cohesive strength favors a low-fracture shear strength and little-to-no permeability increase because dilation associated to slip will be balanced by the comminution, compaction and wearing of the material produced by asperity rupture (Wang et al., 2020). Gutierrez et al. (2000) observe a drastic 5–6 orders of permeability decrease while imposing a 6mm slip on a shale fracture at an effective normal stress of 6 MPa equal or above the fracture asperities uniaxial compressive stress of 4.5-to-5.5 MPa (Fig. 3.2a). This is induced by gouge formation blocking fracture aperture. Nevertheless, they show that fracture permeability after such a decrease remains 3 orders of magnitudes higher than the intact shale permeability. They also observe 3 orders of magnitude increase by shear-induced dilation when the effective normal stress is very low, less than 20% of the initial effective normal stress before shearing.

Fang et al. (2017) confirm that fracture permeability and friction evolution are both driven by the brittleness of asperities, "strong-brittle asperities may result in higher frictional strength, lower frictional stability, and larger permeability than that

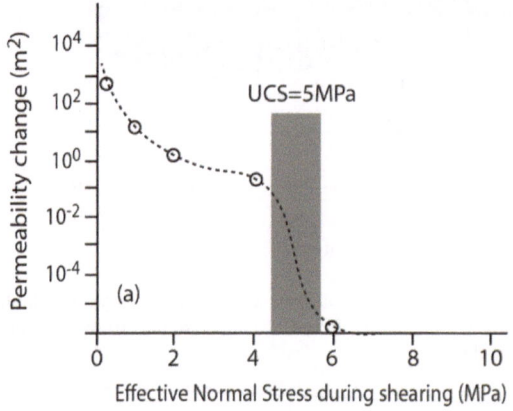

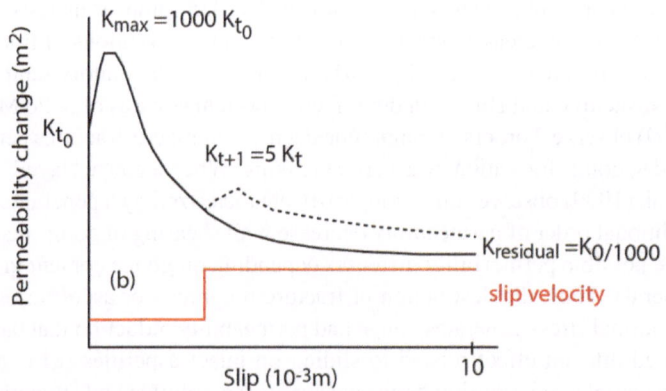

Fig. 3.2 Permeability change at laboratory scale. **a** Permeability-versus-effective normal stress. Data show 5 laboratory shear tests on a single-shale fracture by Gutierrez et al. (2000). For each test, a saw-cut shale cylindrical sample was first confined to a given normal stress then a deviatoric stress loading was applied. Permeability of the fracture was measured before and after the test. Each point in the graph is the difference between these two measurements. **b** Permeability as a function of slip. The effect of an increase of slip velocity is highlighted and marked by an increase of permeability during the global phase of permeability reduction (Fang et al., 2017; Ischibachi et al., 2016, 2018)

of weak-ductile asperities". In a numerical study of fault asperities with a uniaxial compressive strength of 25% lower than the intact rock, Beck et al. (2021) confirm that asperities damage is a key control on fault stability while sliding because of the creation of a low friction wear layer on the fault surface.

3.3 Effect of Slip Velocity and Slip-Rest History (Healing)

Several authors measured in laboratory experiment a 1.05-to-1.6 order of permeability increase after a sudden change in fracture slip velocity (Fig. 3.2b; Fang et al., 2017; Ischibachi et al., 2016, 2018; Ji et al., 2023). Fang et al. (2017) observed that a fast step-increase in sliding velocity generates a sudden increase in fracture permeability. Additionally, Samuelson et al. (2009) observe that the larger the velocity step the larger the permeability change is (with a maximum of about 1.6 order of magnitude). Moreover, Im et al. (2008) report that permeability response to a sliding event is larger if the pre-sliding period is longer. They attribute this behavior to healing of the fracture material associated to time-dependent compaction. The wear material created by asperity destruction during a previous sliding event compacts with time under effective normal stress during a rest period. The longer the period, the larger the compaction that will favor slip-induced dilation at the next sliding event. Thus, since the transient evolution of fault permeability is related to slip velocity, the fluctuation of slip velocity with increasing slip has a major influence, which is modulated by various sizes and strength of surface asperities (Goebel et al., 2012; Xu et al., 2023; Ye and Ghassemi, 2018).

All these laboratory experiments correspond to attempts to relate fracture permeability change with fracture frictional stability (and thus to the potential for induced seismicity). At constant sliding velocity, the decrease in fracture permeability is associated with the increasing number of asperities damage that produce gouge material wearing and finally clogging the fracture (Fig. 3.2b). In the case of a sudden change in slip velocity, the concept is to consider that the time scale is too short to allow for a significant production of wear material. The idea is that the rapid change in slip velocity causes a sudden decrease in the number of contacts between the fracture's walls, in association with a sudden dilation and permeability increase. We may expect that this process also occurs for relatively small total slip displacements. Ishibashi et al. (2018) estimate that it could concern an about 2% asperity contact change. Finally, even these relatively short duration laboratory experiments highlight the importance of healing of the recently wear material, or in other words of the importance of the initial permeability on the magnitude of the permeability increase at failure.

3.4   Effect of Effective Normal Stress

Typical laboratory protocols to explore fracture permeability variation with shear are first to consider either a natural fractured sample, a saw-cut sample or a compacted rock powder to mimic fault gouge. Second, the sample is brought to a given confining stress and its permeability is measured. Third, a deviatoric stress is applied until rupture. Fourth, the test is run until a shear displacement on the order of millimeter is reached in general. Finally, a new permeability test is done. Note that some laboratory devices allow for a "continuous" fracture permeability monitoring.

It is widely recognized that fracture permeability variations are dominated by the variations in effective normal stress (Jaeger et al., 2007). An increase in effective normal stress induces a decrease of both fractures' aperture and dilation during shear. More precisely, the decrease of fracture permeability with stress is usually attributed to a normal closure of the aperture that is related to the deformation of the fracture's asperities (see for example Pyrak-Nolte & Morris, 2000; Raven & Gale, 1985; Rutqvist & Stephansson, 2003; Witherspoon et al., 1980). First, the effective normal stress applied on a fracture is related to the surface of asperities contacting the two opposite fracture walls. For example, Chen et al. (2000) show that effective normal stress increases for two reasons, (1) with the increase in the number of contacts related to the total stress applied on the fracture and (2) with the increase in contacts surface related to erosion and smoothing under fracture slip. Second, the complex asperity geometry, properties and repartition are inducing a local stress heterogeneity that generates fracture permeability and flow field heterogeneity (Kang et al., 2016). Dong et al. (2010) showed the stress dependence is also influenced by the shape of the pores. Schematically, the pores in shale-rich fault zone may be flatter than in quartz-rich fault zone and thus potentially more deformable. They suggest that this impacts whether the permeability variation with stress should be described as a power or as an exponential law.

At high effective normal stress (i.e., normal stress that is close to the fracture normal compressive strength) and under shearing, the crushing of the asperities and the microcracking of the fracture's wall makes the relation between fracture permeability change and stress much more complex. In general, there is an observe fracture permeability decrease but fractures never completely close under high effective normal stresses in the brittle domain (Chen et al., 2000) and at the brittle-ductile transition (Fig. 3.2a; Gutierrez et al., 2000; Petrini et al., 2021). Indeed, after the tests, the fracture permeability may still be 1-to-2 orders of magnitude higher than the intact rock permeability. Moreover, a compilation of fracture-stress permeability curves for different types of rocks shows that there is little variation of fracture permeability for effective normal stress above 5 MPa (Fig. 3.3; Snippe et al., 2022). There is thus an upper and a lower limit to fracture permeability variations under stress variation produced in laboratory-scale experiments.

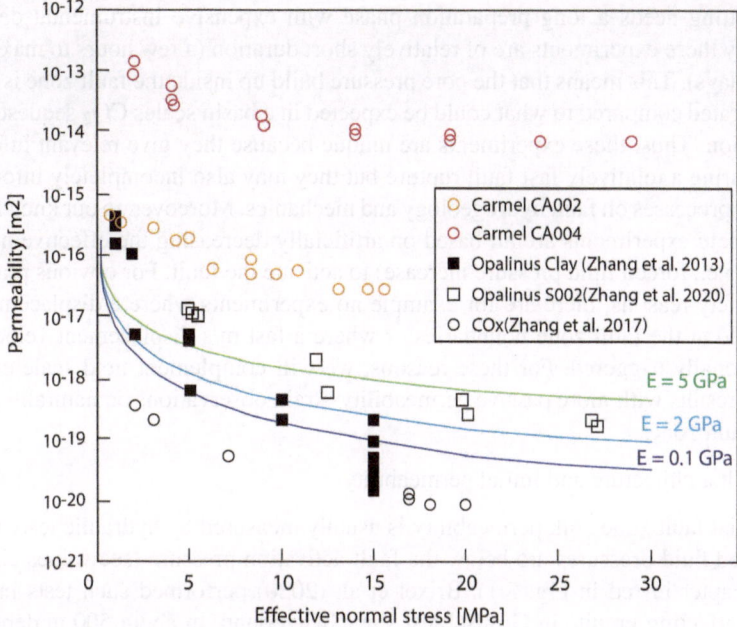

Fig. 3.3 Fracture permeability versus effective normal stress (modified from Snippe et al., 2022). Dots correspond to permeability measurements conducted in the laboratory using gas injections on different argillaceous fractured fault rocks, respectively the Carmel claystone that serves as a caprock for natural CO_2 accumulations in the Paradox Basin (Utah, USA), the Opalinus Clay that is considered as a caprock/hostrock within the Molasse Basin (Switzerland) and the Callovo-oxfordian (Cox) that is considered as a caprock/hostrock in the Paris Basin (France). The curves show an empirical model developed by Snippe et al. (2022) that relate fracture permeability to the intact rock-elastic properties (Young's modulus)

3.5   Fault Zone Compared to Single Fractures or Gouge

The first and best constrained way to study real-world fault activation is through field scale fault activation experiments that allow monitoring fault zone deformations and fluid leakage at the tens of meter scale and at moderate 5–10 MPa confining and deviatoric stresses (Guglielmi et al., 2021). These experiments, representative of natural conditions, are called controlled because the activation is conducted by increasing fault fluid pressure using pressure control or flowrate control protocols. They allow a very high-resolution look into the complex hydromechanical processes that define the fault activation and behavior of slip from nucleation, propagation and arrest, in association with opening. Guglielmi et al. (2020a) observed that complex strain modes in fault affecting clay or carbonate rocks at 350–500 m depths drive fluid leakage under a change in the fault fluid pressure, and eventually produce some moderate and complex micro-seismicity sequence. Field scale experiments are scarce because there are not so many accessible geological fault contexts favorable for testing, and test setting needs a long preparation phase with expensive instrumental devices. Usually these experiments are of relatively short duration (a few hours to maximum a few days). This means that the pore pressure build up inside the fault zone is much accelerated compared to what could be expected in a basin scales CO_2 sequestration operation. Thus, these experiments are unique because they give relevant information during a relatively fast fault rupture but they may also incompletely inform on slower processes on fault hydrogeology and mechanics. Moreover, to our knowledge, these field experiments are all based on artificially decreasing the effective normal stress (i.e., forced fluid pressure increase) to activate the fault. For obvious practical and safety reasons, there are for example no experiments where a displacement is imposed at the fault zone boundaries or where a fast m/s displacement velocity is intentionally triggered. For these reasons, we will complement field scale experiments results with more passive permeability-strain observations on naturally active field fault zones.

- Fault architecture and initial permeability

Initial fault zone bulk permeability is usually measured by hydraulic tests where imposed fluid pressures are below the fault activation pressure (see values cited in this chapter in red in Fig. 3.1). Brixel et al. (2020) performed such tests in fault zones affecting granite in Grimsel test site (Switzerland) at about 500 m depth. At Grimsel, fault zones typically display a core and a damage zone (Caine et al., 1996; Faulkner et al., 2010) with a damage zone permeability of 10^{-12} to 10^{-13} m^2 for fault zones thicknesses ranging from 2 to 10 m (note the fault cores permeability was not measured). Host rock is the least permeable component with values of 10^{-21} to 10^{-18} m^2. This range of fault permeability values consistently spans the range of core values sampled on same type of clay-poor fault zone outcrops (Carpenter et al., 2014; Matsumoto & Shigematsu, 2018; Wibberley & Shimamoto, 2003) and values from hydraulic tests on fault zones affecting some deep geothermal rocks (Guo et al., 2021; Xue et al., 2013). Many studies also show a correlation between permeability (K) and damage zone linear fracture density P_{10} in the form $K \sim P_{10}^a$ with parameter "a" varying from about 1–5 (Bour & Davy, 1997; Brixel et al., 2020; Park et al., 2004 for example). Nevertheless, Brixel et al. (2020) highlight that in addition to fracture density the topology of the damage zone network must be considered to best estimate fault damage zone permeability. They find "high permeability spots" (K > 10^{-14} m^2) at connections between damage zone from different location along, at the tip or between multiple fault cores. Guglielmi et al. (2020) and Nussbaum and Bossart (2004) compiled tests in fault zones affecting the Opalinus clay at the MontTerri underground laboratory at about 350 m depth. Opalinus > 60% clay-rich faults display low bulk permeability of 10^{-20} to 10^{-16} m^2 close to the intact rock permeability of 10^{-18} to 10^{-20} m^2. Guglielmi et al. (2020) suggest that these clay-rich fault zones display a 30-to-80% smaller core to damage zone ratio than fault zones affecting harder rocks that may indicate that fault rupture is more localized in softer rocks. Moreover, these zones contain large lenses of "scaly" fabric which form an anastomosing network of polished surfaces where clay-rich rock splits into progressively smaller flakes (Vannucchi, 2020). Although scaly clay formation is not fully understood, it reveals ductile deformations within the overall brittle fault zone. High clay content, relatively small fracture damage zone and scaly clay lenses are the key factors that explain the low initial permeability of these fault zones affecting typical caprock analogues. Same observations are described by Saffer (2014) from 300 to 600 m deep boreholes drilled through subduction fault zones of Costa Rica and Nankai.

- Natural creep effects on fault zones permeability

Creep is a relatively slow slipping movement (about 1-to-10 cm/year) on a fault that is not producing seismicity (Chen and Bürgmann, 2017), but can influence fault zones permeability. For instance, Saffer (2014) conducted a wide range of direct and indirect permeability observations in ocean drilling project (ODP) boreholes cross-cutting active faults in different subduction zones where displacement rates are around 7–9 cm/year. The conceptual permeability model for such faults is that high permeable 10^{-13} to 10^{-11} m^2 flow paths create in small fraction of the fault surface and spatially migrate or destroy with fault continuously active displacements. Over the long term, the average fault zone permeability stays high around 10^{-16} to 10^{-14} m^2, that is several orders of magnitude higher than the intact sediment.

Barbour (2015) found that the pore pressures response to dynamic strains induced by remote earthquakes in the 100-to-200 m deep boreholes of the Plate Boundary Observatory network depends nonlinearly on the inverse distance perpendicular to actively creeping faults in California. Although there is no fault permeability estimation in this study, Barbour concludes that tectonic creep rates of 1–3 cm/year generate fault zone dilation and associated permeability increase that dissipates fluid pressure. This is caused by dilatant shear on multiple fractures inside the fault zone. Schematically, this may happen as soon as the fault creep rate is higher than the sealing rate.

The sealing or healing rates have also been studied in periods following some large magnitude earthquakes, also called the inter-seismic period. The term "sealing" describes the process of fault permeability decrease. The term "healing" describes the process by which faults regain their strength after an earthquake. Doan et al. (2006) conducted hydraulic tests at 1.1 km depth across the Chelungpu fault zone 6 years after the Mw 7.7 1999 Jiji earthquake. They found a permeability of ~ 10^{-16} m^2 that is a factor of 100 higher than the average intact rock permeability which is a series of silty shales and sandstones. Furthermore, Ma et al. (2019) used water level response to earth tides and repeated hydraulic slug tests in a well intersecting the Qujiang fault (China) at about 200 m depth to monitor fault permeability evolution over a 7 year long period from 2001 to 2018, 31 years after the 1970 Tonghai M 7.8 earthquake that activated the fault. The fault is affecting a thick sandstone layer (intact rock permeability is not given by the authors) and is considered tectonically active with a present-day 4.5 mm/yr dextral strike slip rate (Yu et al., 2020). Although the fault is actively creeping, permeability displays stable values fluctuating of about 25% around 4×10^{-14} m^2 over the 7 years of Ma et al. (2019) study. In addition, no trend was observed highlighting some ongoing sealing of the fault zone following the Main Tonghai earthquake. In both Chelungpu and Qujiang studies, the observed high-fault permeability post-earthquake could suggest, either no sealing, or the occurrence of incomplete sealing with a time scale frame ranging from 6 to 30 years.

Xue et al. (2013) repeated permeability estimations from a 1 year long continuous tidal borehole pressure variations at 800 m depth across the main fault zone, two years after the fault had been activated by the 2008 Mw 7.9 Wenchuan earthquake (China). The tidal approach allows estimating the bulk permeability of a zone radius of about 40 m from the monitoring well, thus integrating a part of the full fault zone thickness estimated to about 100 m. The fault permeability is high, about 1 to 2×10^{-15} m^2, mainly explained by a dense network of hydraulically active fractures in the fault damage zone, and larger than the average intact rock permeability of 1.9×10^{-16} m^2. Over the monitoring period, fault permeability displays a general 4.1×10^{-16} m^2/year decreasing trend that is attributed to post-Wenchuan earthquake healing. In detail, the trend shows fast fluctuations that may correspond to episodic fault movements in response to seismic waves loading from remote earthquakes. Kitagawa et al. (2002) repeated permeability analyses from borehole injections at 540 m depth in the granitic hanging wall fracture damage zone of the Nogima fault zone (Japan) respectively 2 and 5 years after the 1995 Hyogoken-Nanbu Mw 7.3 earthquake that activated the fault. They found a 50% decrease of the permeability in 3 years, corresponding to a decreasing rate of about 1×10^{-13} m^2/year. These two studies show that there may be a quick permeability decrease with time after a large earthquake, of order of months to a few years.

The so-called kinematic permeability, i.e., the permeability decrease with time can be expressed in Eq. (3.1) as a function of the distance from the fault core (x), the depth (z) and the characteristic decay time ($\tau(z)$) (Gratier et al., 2003):

$$K(x, z, t) = \left(\frac{k_0}{\rho g} \right) e^{-x/L} e^{-3t/\tau(z)}, \tag{3.1}$$

where ρ is the fluid density, g is the acceleration of the gravity, k_0 is an initial permeability value, for example before fault activation and L is the half fault thickness. The characteristic decay time $\tau(z)$ depends on the depth in the crust because it is associated to different rates of fault material compaction. Gratier et al. (2003), for example, assumes the predominance of calcite rate in sealing in the low temperature domain and low-pressure upper crust while they assume that quartzite sealing rate dominates at higher temperature and pressure. The depth transition is considered at about 3 km. In Eq. (3.1), the permeability decrease is associated with pressure-solution process occurring at grain scale in the fault. When fitting Eq. (3.1) on permeability data from the San Andreas fault system, Gratier et al. (2003) get large decay times ranging from 70 to > 200 years depending on the initial fault permeability. In Wenchuan study, Xue et al. (2013) finds out that the healing of this fault is relatively fast, corresponding to an exponential decay time of 0.6–2.5 years and invoke that it could reveal that the earthquake induced a pulse of strongly disequilibrated fluids in the fault favoring accelerated crack sealing. It is of minimum several years in the case of the Nojima fault zone (Kitagawa et al., 2002) and potentially in the case of the

Tonghai earthquake fault (Ma et al., 2019). These observations show a high variability in the characteristic time of fault sealing. It is clear that Eq. (3.1) relies on chemical processes spanning over hundreds of years (Gratier, 2011). It is thus also clear that at shorter time scales (below a 100 years) other processes related to the evolution of the rheological and fluid transport properties of fault zones may dominate, and thus end up in potential fault incomplete sealing. For example, the concept from Barbour that a background creeping rate higher than the rate of the chemical processes may maintain a significantly high background fault zone permeability.

- High permeable flow paths creation from field scale fault activation experiments

Based on observations collected during a series of field scale activation experiments in carbonate and shale formations (see Guglielmi et al., 2021), a large permeability increase is observed at the onset of the fault activation phase when cumulated slip is small and slip rate is slow (Fig. 3.4). Hydraulic opening is typically associated with the complex dilatation and shearing of the fractures into the fault zone, both the fault damage zone and the core. The fracture's strain associated with a large fault permeability increase is small of less than a few hundreds of microstrains (Henry et al., 2019). This corresponds to relatively small fractures displacements of 10-to-300×10^{-6} m, marked with normal displacements (u_n) that can be equal or larger than tangential displacements (u_s) ($\delta u_n/\delta u_s$ of 1–1.5 in Guglielmi et al. 2015a, 2020). However, it appears that if for any reasons fault activation stops after this limited amount of strain, there is a lot of reversible strain, as well as local back shear on fractures (Henry et al., 2019). For instance, for faults in shales, rupture on the main fault plane is triggered after several hours of injection. It corresponds to a much larger amount of fault shear slip compared to dilation, respectively $\delta u_n/\delta u_s \ll 1$. Most of the deformations remain permanent and there is less-to-no permeability increase during this period, i.e., injection flowrate gets to an almost steady state. In the different field experiments (Guglielmi et al., 2020), this transition occurs after a cumulative amount of slip of about 300-to-400 $\times 10^{-6}$ m.

In these field experiments, a permeability increase of several orders of magnitude is observed during the slow slip phase (Cappa et al., 2018; Guglielmi et al., 2015b). Below 400×10^{-6} m, the large permeability increase is controlled by the connection between different fractures within the fault zone. Donze et al. (2020) matched the nonlinear permeability variation observed in the Tournemire experiment (France) by summing the effect of a few discrete fractures (soft behavior) that have a high permeability variation under low effective stress variation. Additionally, the effect of bulk micro-fractured volumes (hard behavior) within the fault zone displays lower permeability variations related to changes in the tortuosity of the micro-fractures network (Donze et al., 2020 adapted the concept developed by Zheng et al., 2015). They show that the soft behavior can eventually initiate a significant local fault permeability increase, thus that some leakage flow paths may be developed under elastic changes in effective stress thus a globally inactive fault. The amount of soft versus hard behavior and the initial bulk fault permeability may characterize the difference in permeability change between clay-rich faults compared to less than 20% clay

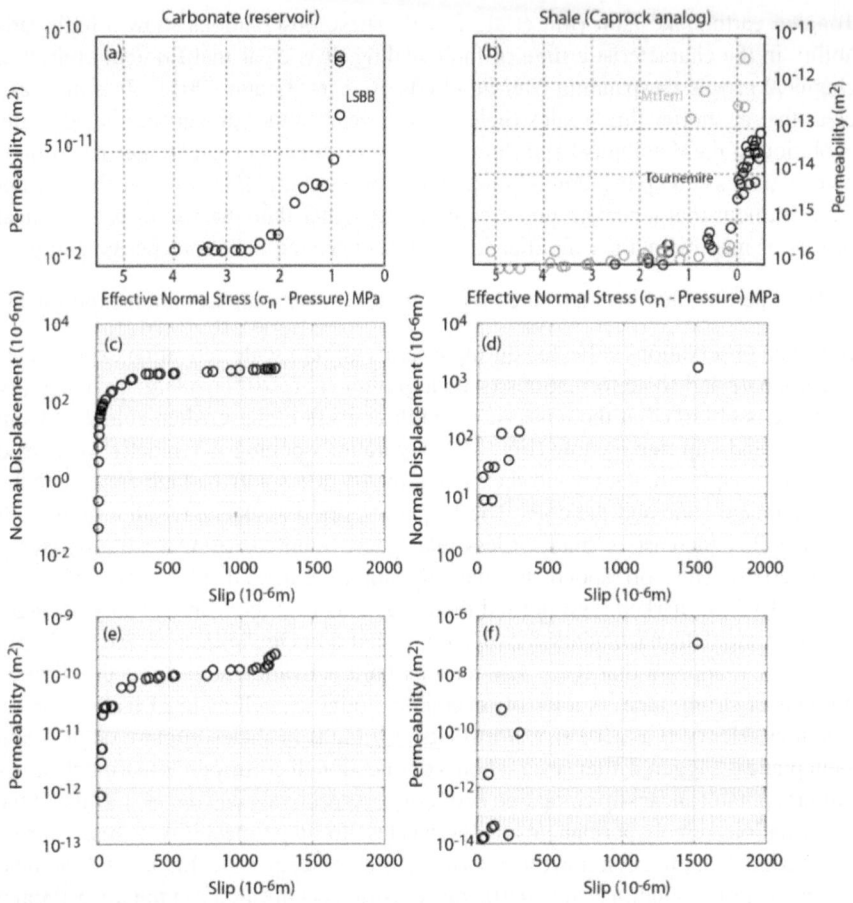

Fig. 3.4 Fault zone permeability variations observed in field scale experiments in a carbonate fault (left column) and in a shale fault (right column). **a** and **b** Permeability variation with fault effective normal stress. **c** and **d** Fault normal versus slip displacements. **e** and **f** Permeability variation with fault slip

faults. Cappa et al. (2022a, 2022b) also observed a 2–3 orders permeability increase associated to a hundred micrometers dilatant fault slip affecting a porous carbonate series during a field scale fault activation experiment at about 380m depth in the LSBB underground research laboratory (France). Guglielmi et al. (2021) observed up to 5 orders of permeability increase associated to 10–100 µm fault slip affecting the Opalinus clay (considered as a caprock analogue) at 370 m depth in the Mont Terri research laboratory (Switzerland).

At the same time, recent experiments have demonstrated that local flow path(s) develop within a much larger slow rupture fault patch. Considering the Mont Terri fault as a single frictional interface, Cappa et al. (2022a, 2022b) studied the spatio-temporal decoupling between slip and opening along the interface. They reproduce the observed growth of a localized high-pressure leakage flow path from a source point in the fault that represents the field injection borehole. They also show that a rupture area synchronously develops beyond the pressurized area, where dilatant slip pre-opens the fault for fluid leakage to propagate. In this zone, pore pressure tends to slightly decrease because of slip-induced dilation and slip velocities are slow of 0.1 to 1 μm/s. This behavior beyond the pressurized zone creates favorable conditions for induced seismicity on fault asperities ready to slip seismically before injection. In detail within the pressurized zone, high dilation and slip velocities > 10 μm/s create favorable conditions for developing aseismic slip. Thus, there is also a decoupling between the localization of fluid leakage and the localization of seismic events, most of them being outside the leakage flow path. Moreover, this concept explains potentially induced seismicity in clay-rich faults where slip is usually considered mostly aseismic due to rate-strengthening frictional properties over the fault. This depressurized slow slipping zone is at least twice or more larger than the pressurized flow path. This may explain why methods that use the propagation of seismic events as proxies to fluid diffusion front end up overestimating fault permeability (Shapiro et al., 1997). At present, the sensitivity of the fault permeability increase to this spatial–temporal decoupling between slip and opening has yet to be refined but these results highlight a large influence of aseismic slow movements on fluid leakage in faults and on the triggering of induced seismicity. Recently, Shadoan et al. (2023) produced time lapse images of a leakage flow path growth during injection experiments in the Mont Terri fault. These images confirm the analyses from Cappa et al. (2022a, 2022b). First, a localized flow path develops following some complex fault zone heterogeneities. Second, a dilatant slip zone propagates in front of the leakage flow path as seen on monitoring holes clearly localized outside the flow path image. Third, seismicity is located either at the front of the propagating flow path, or several meters away.

Furthermore, experiments show that a major fault strain/slip localization may occur after a period of bulk viscoplastic strain and associated limited fluid leakage, and result in a significant fault zone bulk permeability increase. These field scale experiments also allow exploring how permeability varies within the thickness of the fault during its activation. In the Tournemire field experiments (France), Henry et al. (2019) used two boreholes crosscutting the entire 6-to-8 m thick fault zone and equipped respectively with strain gauges distributed every 0.5-to-1 m and with electric resistivity electrodes distributed every 0.3 m across the fault zone. The fault zone contains a 1-to-2 m thick core surrounded on both sides by a 2-to-3 m thick fracture damage zone. In the experiment described here, half of the fault zone including the core and the hanging wall damage zone was isolated with a straddle packer system in a third borehole dedicated to fluid injection. Then a constant pressure fluid injection was conducted for 12 h while continuously monitoring strain and electric resistivity variations distributed within the fault zone thickness. This experiment shows that fault starts leaking at three fractures located in the damage zone at the edges of the core zone. Almost 3 h later, there is a major slip event at the core. During this event, flow rate increases by a factor of 4, and new fractures leakage are activated, some being in the footwall damage zone.

Permeability evolution during the nucleation phase of fault slip depends on the slip rate and states of the activated fractures within the fault zone volume. In addition, Jeanne et al. (2018) show that during the period of fault slip initiation corresponding to cumulated slip below 400×10^{-6} m and slow slip velocities of 0.5–500 μm/s in the studied MontTerri fault activation experiments, the permeability depends on the local strain rates and on the evolution of fractures state in addition to pressure. Quasi-similar fault movements can create or destroy the local fluid flow paths depending on the heterogeneity that is activated within the fault zone, and if a flow path is activated it modifies the fault pore pressure and impacts the nucleation process of fault slip. For example, they observe that a too large local flow path permeability increase will slow down the propagation of secondary ruptures within the fault zone. Interestingly, Cappa et al. (2022a, 2022b) also find that a permeability model associated with rate-and-state friction best fits the permeability evolution during the fault slip nucleation period in the case of an initially permeable carbonate fault experiment. Moreover, they show that almost 60% of permeability evolution may occur after cumulated fault slip less than 120×10^{-6} m.

- Potential fault self-sealing mechanism from a clay-rich fault field activation experiment

Generally, only a partial sealing of the activated fault zone is observed following a field scale activation experiment. Guglielmi et al. (in press, 2024) conducted a series of hydraulic pulse tests to monitor the long-term permeability evolution following a fault activation experiment at Mont Terri (Fig. 3.5a). Pressure pulse tests were repeated every 2 weeks during the year following the fault activation event. Fault permeability variations were compared to fault movements monitored during and after activation by a multimodal network of distributed optical fibers (DSS) and local three-dimensional borehole displacement sensors. The pre-activation fault permeability was estimated to 10^{-16} m^2 while the surrounding intact rock permeability is 10^{-18} to 10^{-20} m^2. There is a five orders of magnitude permeability increase to a maximum of 3.2×10^{-11} m^2 during fault activation associated to a 600×10^{-6} m fault displacement (Fig. 3.5b, c).

Early post-activation, transient compaction and the subsequent slow compaction were observed, transitioning to an extension regime. The pulse tests showed a fast permeability drop by more than two orders of magnitude within the first 10 days, followed by a gradual and less pronounced decrease. During the 70 days post-activation, this slow fault permeability decrease is thus dominated by a slow ~ 1.6×10^{-11} s^{-1} fault creep. After 70 days, additional factors, such as clay mineral swelling are hypothesized to become the dominant mechanism of permeability decrease. After one year of post-activation monitoring, the fault permeability is still three orders of magnitude higher than before the activation, highlighting an incomplete sealing. A creep-dependent permeability law may explain such a relatively short but incomplete permeability decrease with time. Extrapolation suggested a sealing process taking at least 50 years.

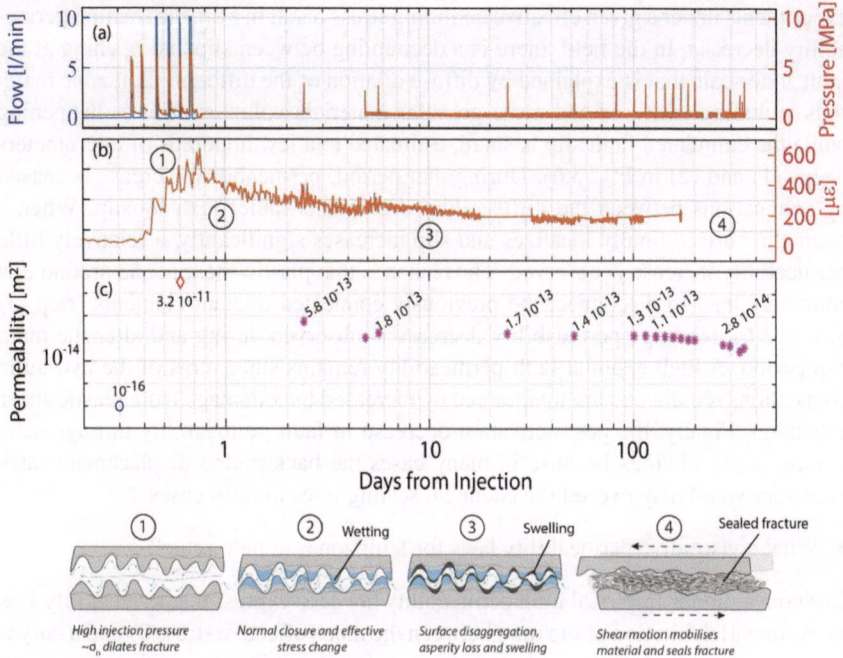

Fig. 3.5 Long-term clay fault hydromechanical response. **a** Flow rate injected in the fault zone and fault pore pressure versus time (pulse tests correspond to pressure spikes from 2.3 days after activation end); **b** Fault strain measured versus time (strain was measured along a borehole drilled roughly perpendicular to the fault surface); **c** Estimated values of fault hydraulic transmissivity as a function of time. Drawings 1–4 are modified from Fang et al. (2017). Drawings schematically figure the evolution of a rough clay-rich fault under shearing and swelling

3.6 Conclusion—Coupling Fault Zone Permeability to Rupture

- Conceptual fault permeability variation from field experiments

Field observations show some typical characteristics of fault zones permeability change at activation compared to laboratory-scale experiments that often focus on single-fault deformation types such as gouge or single-fracture behavior while in the field a fault zone integrates multiscale features (step (0) in Fig. 3.6). What is common at field and laboratory scales is the strong dependency of permeability on effective stress. Fault zone permeability decreases with increasing effective stress with about the same power law dependency of permeability on stress. What is significantly different is the way permeability varies with fault opening and slip. In the laboratory, the dilation of a single fracture is induced by slip on the fracture asperities until a given amount of slip where asperities start to destroy, producing material that ends up clogging almost completely the fracture. Most laboratory fracture shear experiments under a given effective confining stress result in a drastic fracture permeability decrease. In the field, there is a decoupling between slip and opening at the fault zone scale that is explained by diffuse dilation of the different fault zone materials including fractures and more granular materials volumes. This is happening while the cumulated fault slip is small, estimated to a few hundreds of micrometers (steps (1) and (2) in Fig. 3.6). During that period, permeability increase is caused by connections between the diffuse-dilated elements rather than by slip. When a main slip surface finally localizes and slip increases significantly, a relatively little permeability increase is observed. The reason is that plastic shear at and around this zone may tend to disconnect the previously connected dilatant elements (step (3) in Fig. 3.6). No large permeability decrease is observed during and after the main slip period. A high residual fault permeability remains since most of the hydraulic connections remain and are maintained by increased bulk damage from plastic shear activation. Finally, the post-activation decrease in fault permeability through self-sealing is not obvious because in many cases the background displacement rates even very small may exceed the chemical sealing rates in most cases.

- What constitutive permeability laws for fault zones at field scale?

One conventional empirical fault permeability law is to express the permeability k as an exponential function of the ratio between the mean effective stress $\overline{\sigma}\prime$ and a curve-fitting parameter σ^* figuring the sensitivity of permeability to confining pressure

Fig. 3.6 Fault zone permeability evolution with fault slip from field scale experiments and at a given initial state of stress. Upper graph shows the fault bulk deformation versus the fault cumulated slip. The lower graph shows the fault permeability change versus slip. Steps (0) to (3) are explained in the text

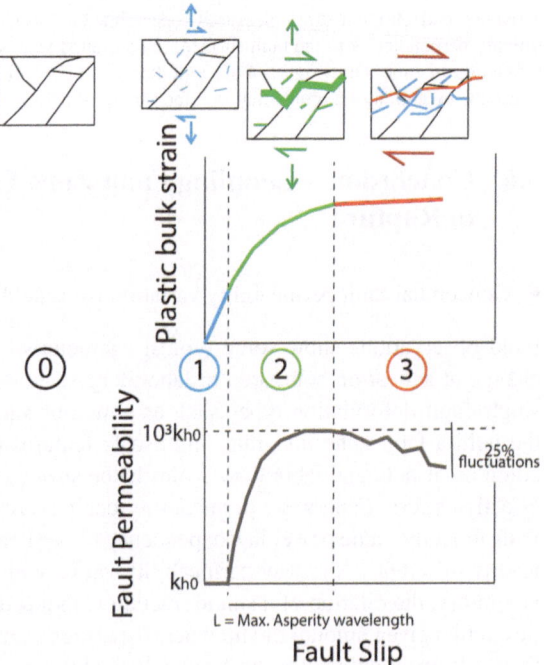

(Rice, 1992):

$$k = k_{min} + (k_0 - k_{min})e^{\left(\frac{\overline{\sigma}\prime}{\sigma*}\right)}, \tag{3.2}$$

where k_0 is originally the permeability at zero effective stress and, it is considered the maximum fault permeability. If we refer to Fig. 3.1, we could consider a value around 10^{-12} m^2 for k_0. Ishii (2015) empirically related the maximum fault zone permeability to the rock ductility index DI defined as $DI = \frac{\overline{\sigma}\prime}{\sigma_t}$, where σ_t is the tensile strength of the intact rock surrounding the fault zone. Looking at field data from 6 underground laboratories worldwide nested in different hard to soft rock types, he suggested a maximum permeability of 10^{-13} m^2 for DI $\ll$ 1 corresponding highly brittle hard rock faults. He also suggested that clay-rich faults with DI > 4–5 may have a k_0 very low and close to the intact rock.

k_{min} is the minimum background permeability. The parameter $\sigma*$ that characterizes the fault zone permeability sensitivity to confining pressure is typically ranging between 5 and 30 MPa (Evans et al., 1997; Marguin & Simpson, 2023). It depends on the rock type with values of 30–40 MPa for intact hard rock such as gneiss-granite (Bernabe, 1986; David et al., 1994; Morrow et al., 1984), and values of 9–20 MPa for fractured hard rock or clay-rich fault gouge (Evans et al., 1997; Kranz et al., 1979). David et al. (1994) suggests that permeability reduction that occurs in samples with large values of $\sigma*$ may result from crack closure.

A second type of law comes from laboratory experiments on porous media permeability variation with compaction. Basically, the fault zone is considered as an equivalent porous media where the porosity exponentially depends on the effective mean stress (Shi & Wang, 1986):

$$\emptyset = \emptyset_0 e^{(-\beta\overline{\sigma}\prime)}, \tag{3.3}$$

where β is a material constant.

Fault permeability is related to porosity change through a more or less complex power law such as the one in Eq. 3.4:

$$\frac{k}{k_0} = \left(\frac{\emptyset}{\emptyset_0}\right)^n, \tag{3.4}$$

where k_0 and $\emptyset_0$ are reference permeability and porosity, and n is a constant that can vary from 1 to 25, depending on rock type and state of stress (for a review see Yang & Dunham, 2021).

Then, one example relationship between permeability, porosity and effective stress was given by Zheng et al. (2015) based on the conceptual idea that, during mechanical compaction, high deformable cracks first close at low stresses and then intergranular pores deform at higher stress. An analogue to a fault zone would be compliant flat fractures closing at low stress variation followed by closure of more "spherical" microcracks in the granular damaged volumes between fractures at higher stresses as was applied by Donze et al. (2020) on field scale shale fault at Tournemire (France). The permeability law takes the form of a two-component law, respectively a soft and a hard component:

$$k = k_{e,0}e^{-\beta C_e \varnothing_{e,0} \bar{\sigma}\prime} + \alpha \left[\gamma_t e^{\left(-\frac{\bar{\sigma}\prime}{K_t}\right)} \right]^m, \tag{3.5}$$

where the first term describes the permeability change of the hard component with β, C_e, $\varnothing_{e,0}$, $k_{e,0}$ being a material parameter, the compliance, the initial porosity and the initial permeability at zero confining stress of the hard part component, respectively. The second term describes the permeability change of the soft component with α and m being material constant. γ_t is the ratio between the soft volume of the rock at a given stress versus the unstress soft volume. K_t is the permeability of the soft volume. In this relationship, we may infer that K_t (that figures fractures in a fault zone) can be described based on the analogy of flow between two parallel plates, which is frequently called cubic law (Zimmerman & Bodvarsson, 1996). Zheng et al. (2015) and Donze et al. (2020) successfully matched this soft-hard component permeability law on permeability versus stress data from laboratory tests on shale samples (Dong et al., 2010) and on permeability estimated from a shale fault field scale activation, respectively.

These two types of stress-dependent permeability laws may be well-suited to describe the highly dilatant fault zone at the onset of activation, while the amount of slip is relatively small (less than the single fault fractures average asperity wavelength as suggested in Fig. 3.4). Nevertheless, these laws take little consideration of large slip damage associated with moderate-to-isochoric fault zone volume change.

To describe the permeability change at large slip, some permeability laws integrate the plastic evolution of the fault zone at rupture in addition to the change in effective stress. Hsiung et al. (2005) and Rinaldi et al. (2015) describe a fault zone as a transversely isotropic layer characterized by a single fracture set parallel to the fault dip direction. They suggest that the change in fault permeability depends on the change in the effective normal stress applied to the fault zone, on the plastic shear e_{pss} and on the tensile strength e_{pts} of fractures. Jeanne et al. (2018) used this permeability law to match permeability changes of the Mont Terri shale fault that was activated by controlled fluid injections. The relationship is like a complex "cubic law" writing as follows:

$$k_{hm} = k_0 \left[\frac{a}{c(c\sigma_n\prime + 1)} \sqrt{\frac{\varnothing_0}{12k_0}} + \frac{\varnothing_{hm} - \varnothing_0}{\varnothing_0} \right]^3$$

$$= k_0 \left[\frac{a}{c(c\sigma_n\prime + 1)} \sqrt{\frac{\varnothing_0}{12k_0}} + \frac{e_{pts} + e_{pss}\tan\varphi}{\varnothing_0} \right]^3, \tag{3.6}$$

where $a = K^{-1}$ and K is the fractures normal stiffness, and

$$c = \frac{-1 \pm \sqrt{1 + 4\sigma_n{}'a\sqrt{\frac{\varnothing_0}{12k_0}}}}{2\sigma_n 0'}. \tag{3.7}$$

The last term of the equation describes plastic porosity changes according to

$$\varnothing_{hm} - \varnothing_0 = e_{pts} + e_{pss} \tan \varphi, \tag{3.8}$$

where φ is the fractures dilation angle.

Some variants of the above law integrate that the porosity may change with the velocity of fault plastic shear displacement (Jeanne et al., 2018). In this case, the porosity change follows a rate-and-state dependency on velocity based on Segall and Rice (1995) equation:

$$\varnothing_{hm} - \varnothing_0 = -\varepsilon \ln\left(\frac{\theta V_0}{D_c}\right) \tag{3.9}$$

Using this last equation allows consideration of the effects of both slip rate increase and decrease on fault permeability change. In an indirect way, they allow to figure the creation or destruction of flow paths in a moving fault zone. This type of law describes permeability variation at the main localized shear zone within the activated fault zone through complex adjustments made to the basic concept of hydraulic aperture. It thus neglects description of permeability in the remaining bulk volume of the fault zone. To account for fault zone permeability outside the main shear zone, some variants suggest adding a tortuosity factor (Fang et al., 2017). Another solution could be to combine the hard permeability term of Zheng et al. (2015) with a soft parameter as described by Jeanne et al. (2018).

References

Barbour, A. J. (2015). Pore pressuresensitivities to dynamic strains: Observations in active tec-tonic regions. *Journal of Geophysics Research: Solid Earth, 120*, 5863–5883. https://doi.org/10.1002/2015JB012201

Barton, N., Bandis, S., & Bakhtar, K. (1985). Strength, deformation and conductivity coupling of rock joints. *International Journal of Rock Mechanics and Mining Sciences, 22*(3), 121–140.

Beck, J. M., Mair, K., & Renard, F. (2021). Decrypting healed fault zones: How gouge production reduces the influence of fault roughness. *Geophysical Journal International, 225*, 759–774.

Bernabe, Y. (1986). The effective pressure law for permeability in Chelmsford granite and Barre granite. *International Journal of Rock Mechanics Mineral Sciences and Geomechanical Abstracts, 23*, 267–275.

Bour, O., & Davy, P. (1997). Connectivity of random fault networks following a power law fault length distribution. *Water Resources Research, 33*(7), 1567–1583.

Brixel, B., Klepikova, M., Jalali, M.R., Lei, Q., Roques, C., Kriestch, H., & Loew, S. (2020). Tracking fluid flow in Shallow crustal fault zones: 1. Insights from single-hole permeability estimates. *Journal of Geophysical Research: Solid Earth, 125.*

Caine, J. S., Evans, J. P., & Forster, C. B. (1996). Fault zone architecture and permeability structure. *Geology, 24*(11), 1025–1028

Cappa, F., Guglielmi, Y., & de Barros, L. (2022a). Transient evolution of permeability and friction in a slowly slipping fault activated by fluid pressurization. *Nature Communications, 13,* 3039. https://doi.org/10.1038/s41467-022-30798-3

Cappa, F., Guglielmi, Y., Nussbaum, C., & Birkholzer, J. (2018). On the relationship between fault permeability increases, induced stress perturbation, and the growth of aseismic slip during fluid injection. *Geophysical Research Letters, 45,* 11,012–11,020. https://doi.org/10.1029/2018GL 080233

Cappa, F., Guglielmi, Y., Nussbaum, C., De Barros, L., & Birkholzer, J. (2022b). Fluid migration in low-permeability faults driven by decoupling of fault slip and opening. *Nature Geosciences.* https://doi.org/10.1038/s41561-022-00993-4

Carpenter, B. M., Kitajima, H., Sutherland, R., Townend, J., Toy, V. G., & Safer, D. M. (2014). Hydraulic and acoustic properties of the active Alpine Fault, New Zealand: Laboratory measurements on DFDP-1 drill core. *Earth and Planetary Science Letters, 390,* 45–51. https://doi.org/10.1016/j.epsl.2013.12.023

Chen, K. H., & Bürgmann, R. (2017). Creeping faults: Good news, bad news? *Reviews of Geophysics, 55,* 282–286. https://doi.org/10.1002/2017RG000565

Chen, Z., Narayan, S. P., Yang, Z., & Rahman, S. S. (2000). An experimental investigation of hydraulic behavior of fractures and joints in granitic rock. *International Journal of Rock Mechanics and Mining Sciences, 37*(7), 1061–1071. https://doi.org/10.1016/S1365-160 9(00)00039-3

David, C., Wong, T., Zhu, W., & Zhang, J. (1994). Laboratory measurement of compaction-induced permeability change in porous rocks: Implications for the generation and maintenance of pore pressure excess in the crust. *Pure and Applied Geophysics, 143,* 425–456.

Doan, M. L., Brodsky, E. E., Kano, Y., & Ma, K. F. (2006). Insitu measurement of the hydraulic diffusivity of the active Chelungpu Fault, Taiwan. *Geophysical Research Letters, 33,* L16317. https://doi.org/10.1029/2006GL026889

Dong, J. J., Hsu, J. Y., Wu, W. J., Shimamoto, T., Hung, J. H., Yeh, E. C., Wu, Y. H., & Sone, H. (2010). Stress-dependence of permeability and porosity of sandstone and shale from TCDP Hole-A. *International Journal of Rock Mechanics & Mining Sciences, 47,* 1141–1157.

Donze, F. V., Tsopela, A., Guglielmi, Y., Henry, P., & Gout, C. (2020). Fluid migration in faulted shale rocks: Channeling below active faulting threshold. *Europen Journal of Environmental and Civil Engineering.* https://doi.org/10.1080/19648189.2020.1765200

Evans, J. P., Forster, C. B., & Goddard, J. V. (1997). Permeability of fault-related rocks, and implications for hydraulic structure of fault zones. *Journal of Structural Geology, 19*(11), 1393–1404. https://doi.org/10.1016/s0191-8141(97)00057-6

Fang, Y., Elsworth, D., Wang, C., Ishibashi, T., & Fitts, J. P. (2017). Frictional stability-permeability relationships for fractures in shales. *Journal of Geophysics Research: Solid Earth, 122,* 1760–1776.

Faoro, L. A., Niemeijer, A., Marone, C., & Elsworth, D. (2009). Influence of shear and deviatoric stress on the evolution of permeability in fractured rock. *Journal of Geophysical Research, 114*(1).

Farrell, N. J. C., Debenham, N., Wilson, L., Wilson, M. J., Healy, D., King, R. C., Holford, S. P., & Taylor, C. W. (2021). The effect of authigenic clays on fault zone permeability. *Journal of Geophysical Research: Solid Earth, 126.*

Faulkner, D., Jackson, C., Lunn, R., Schlische, R., Shipton, Z., Wibberley, C., & Withjack, M. (2010). A review of recent developments concerning the structure, mechanics and fluid flow properties of fault zones. *Journal of Structural Geology, 32,* 1557–1575. https://doi.org/10.1016/j.jsg.2010.06.009

Goebel, T. H. W., Becker, T. W., Schorlemmer, D., Stanchits, S., Sammis, C., Rybacki, E., & Dresen, G. (2012). Identifying fault heterogeneity through mapping spatial anomalies in acoustic emission statistics. *Journal of Geophysical Research, 117*(B3), B03310. https://doi.org/10.1029/2011JB008763

Gratier, J. P. (2011). Fault permeability and strength evolution related to fracturing and healing episodic processes (years to millennia): The role of pressure solution. *Oil & Gas Science and Technology—Revue De l IFP, 3*(3), 491–506. https://doi.org/10.2516/ogst/2010014

Gratier, J.-P., Favreau, P., & Renard, F. (2003). Modeling fluid transfer along California faults when integrating pressure solution crack sealing and compaction processes. *Journal of Geophysical Research, 108*(B2), 2104. https://doi.org/10.1029/2001JB000380

Guglielmi, Y., Elsworth, D., Cappa, F., Henry, P., Gout, C., Dick, P., & Durand, J. (2015a). In situ observations on the coupling between hydraulic diffusivity and displacements during fault reactivation in shales. *Journal of Geophysical Research: Solid Earth, 120*, 7729–7748. https://doi.org/10.1002/2015JB012158

Guglielmi, Y., Elsworth, D., Cappa, F., Henry, P., Gout, C., Dick, P., & Durand, J. (2015b). In situ observations on the coupling between hydraulic diffusivity and displacements during fault reactivation in shales. *Journal of Geophysical Research: Solid Earth, 120*, 7729–7748. https://doi.org/10.1002/2015JB012158

Guglielmi, Y., Nussbaum, C., Jeanne, P., Rutqvist, J., Cappa, F.,& Birkholzer, J. (2020a). Complexity of fault rupture and fluid leakage in shale: insights from a controlled fault activation experiment. *Journal of Geophysical Research: Solid Earth, 125*, e2019JB017781. https://doi.org/10.1029/2019JB017781

Guglielmi, Y., Nussbaum, C., Cappa, F., DeBarros, L., Rutqvist, J., & Birkholzer, J. (2021). Field-scale fault reactivation experiments by fluid injection highlight aseismic leakage in caprock analogs: Implications for CO_2 sequestration. *International Journal of Greenhouse Gas Control, 111*, 103471.

Guglielmi, Y., Nussbaum, C., Jeanne, P., Rutqvist, J., Cappa, F., & Birkholzer, J. (2020). Complexity of fault rupture and fluid leakage in shale: Insights from a controlled fault activation experiment. *Journal of Geophysical Research.* https://doi.org/10.1029/2019JB017781e2019JB017781

Guo, H., Brodsky, E. E., Goebel, T. H. W., & Cladouhos, T. T. (2021). Measuring fault zone and host rock hydraulic properties using tidal responses. *Geophysical Research Letters, 48*, e2021GL093986.

Gutierrez, M., Øino, L. E., & Nygård, R. (2000). Stress-dependent permeability of a de-mineralised fracture in shale. *Marine and Petroleum Geology, 17*, 895–907.

Henry, P., Guglielmi, Y., Gout, C., Castilla, R., Dick, P., Donze, F., Tsopela, A., Neyens, D., DeBarros, L., & Durand, J. (2019). Strain and flow pathways in a shale fault zone: An In-situ test of fault seal integrity. In *Fifth International Conference on Fault and Top Seals 2019, EAGE*, Sep. 2019, Palermo, Italy.

Hsiung, S. M., Chowdhury, A. H., & Nataraja, M. S. (2005). Numerical simulation of thermal-mechanical processes observed at the drift-scale heater test at Yucca Mountain, Nevada, USA. *International Journal of Rock Mechanics and Mining Science, 42*, 652–666.

Im, K., Elsworth, D., & Fang, Y. (2018). The influence of pre slip sealing on the permeability evolution of fractures and faults. *Geophysical Research Letters, 45*, 166–175. https://doi.org/10.1002/2017GL076216

Ishibashi, T., Elsworth, D., Fang, Y., Riviere, J., Madara, B., Asanuma, H., Watanabe, N., & Marone, C. (2018). Friction-stability-permeability evolution of a fracture in granite. *Water Resources Research, 54*(12), 9901–9918.

Ishibashi, T., Watanabe, N., Asanuma, H., & Tsuchiya, N. (2016). Linking microearthquakes to fracture permeability change: The role of surface roughness. *Geophysical Research Letters, 43*, 7486–7493. https://doi.org/10.1002/2016GL069478

Ishii, E. (2015). Predictions of the highest potential transmissivity of fractures in fault zones fron rock rheology: Preliminary results. *Journal of Geophysics Research: Solid Earth, 120*.

Jaeger, J. C., Cook, N. G. W., & Zimmerman, R. W. (2007). *Fundamentals of rock mechanics.* Wiley-Blackwell.

Jeanne, P., Guglielmi, Y., Rutqvist, J., Nussbaum, C., & Birkholzer, J. (2018). Permeability variations associated with fault reactivation in a claystone formation investigated by field experiments and numerical simulations. *Journal of Geophysical Research: Solid Earth, 123,* 1694–1710.

Ji, Y., Zhang, W., Hofmann, H., Cappa, F., & Zhang, S. (2023). Fracture permeability enhancement during fluid injection modulated by pressurization rate and surface asperities. *Geophysical Research Letters, 50,* e2023GL104662. https://doi.org/10.1029/2023GL104662

Kang, P. K., Brown, S., & Juanes, R. (2016). Emergence of anomalous transport in stressed rough fractures. *Earth and Planetary Science Letters, 454,* 46–54.

Kitagawa, Y., Fujimori, K., & Koizumi, N. (2002). Temporal change in permeability of the rock estimated from repeated water injection experiments near the Nojima fault in Awaji Island, Japan. *Geophysical Research Letters, 29*(10), 1483. https://doi.org/10.1029/2001GL014030

Kranz, R. L., Frankel, A. D., Engelder, T., & Scholz, C. H. (1979) The permeability of whole and jointed Barre granite. *Internutionul Journal Rock Mechanics Mineral Science and Gromrchanicul Abstracts, 16,* 225234.

Li, Z., Ma, X., Kong, X. Z., Saar, M. O., & Vogler, D. (2023). Permeability evolution during pressure-controlled shear slip in saw-cut and natural granite fractures. *Rock Mechanics Bulleting, 2,* 100027.

Ma, Y., Wang, G., Yan, R., & Wang, B. (2019). Long-term in situ permeability variations of an active fault zone in interseismic period. *Pure Applied Geophysics, 176,* 5279–5289.

Marguin, V., & Simpson, G. (2023). Influence of fluids on earthquakes based on numerical modeling. *Journal of Geophysical Research: Solid Earth, 128,* e2022JB025132. https://doi.org/10.1029/2022JB025132

Matsumoto and Shigematsu Earth. (2018). *Planets and Space, 70,* 13. https://doi.org/10.1186/s40623-017-0765-5

Morrow, C. A., Shi, L. Q., & Byerlee, J. D. (1984). Permeability 01 fault gouge under confining pressure and shear stress. *Journal of Geophysical Research, 89,* 3193–3200.

Nussbaum, C., & Bossart, P. (2004). Compilation of K-values from packer tests in the Mont Terri rock laboratory. Mont Terri Technical Note, TN 2005-10, p. 29. Federal Office of Topography (swisstopo), Wabern, Switzerland.

Park, Y., Sudicky, E. A., Mclaren, R. G., & Sykes, J. F. (2004). Analysis of hydraulic and tracer response tests within moderatelyfractured rock based on a transition probability geostatistical approach. *Water Resources Research, 40,* W12404.

Petrini, C., Madonna, C., & Gerya, T. (2021). Inversion in the permeability evolution of deforming Westerly granite near the brittle–ductile transition. *Science Reports, 11,* 24027 (2021). https://doi.org/10.1038/s41598-021-03435-0

Pyrak-Nolte, L. J., & Morris, J. P. (2000). Single fractures under normal stress: The relation between fracture specific stiffness and fluid flow. *International Journal of Rock Mechanics and Mining Sciences, 37*(1–2), 245–262.

Raven, K. G., & Gale, J. E. (1985). Water flow in a natural rock fracture as a function of stress and sample size, Elsevier. *International Journal of Rock Mechanics and Mining Geomechanics Abstracts, 22,* 251–261.

Rice, J. R. (1992). Fault stress states, pore pressure distributions, and the weakness of the San Andreas fault. *International Geophysics, 51,* 475–503.

Rinaldi, A. P., Rutqvist, J., Sonnenthal, E. L., & Cladouhos, T. T. (2015). Coupled THM modeling of hydroshearing stimulation in tight fractured volcanic rock. *Transport in Porous Media, 108*(1), 131–150. https://doi.org/10.1007/s11242-014-0296-5

Rutqvist, J., & Stephansson, O. (2003). The role of hydromechanical coupling in fractured rock engineering. *Hydrogeology Journal, 11*(1), 7–40.

Saffer, D. (2014). The permeability of active subduction plate boundary faults. *Geofluids, 2015*(15), 193–215. https://doi.org/10.1111/gfl.12103

Samuelson, J., Elsworth, D., & Marone, C. (2009). Shear-induced dilatancy of fluid-saturated faults: Experiment and theory. *Journal of Geophysical Research, 114*, B12404. https://doi.org/10.1029/2008JB006273

Segall, P., & Rice, J. R. (1995). Dilatancy, compaction, and slip instability of a fluid infiltrated fault. *Journal of Geophysical Research: Solid Earth, 100*, 22,155–22,171.

Shadoan, T. A., Ajo-Franklin, J. B.,Guglielmi, Y., Wood, T., Robertson, M., Cook, P., et al. (2023). Active-sourceseismic imaging of fault re-activationand leakage: An injection experimentat the Mt Terri Rock Laboratory, Switzerland. *Geophysical Research Letters, 50*, e2023GL104080. https://doi.org/10.1029/2023GL104080

Shapiro, S.A., Huenges, E., & Borm, G. (1997). Estimating the crust permeability from fluid-injection-induced seismic emission at the KTB site. *Geophysical Journal International, 131*(2), F15–F18. https://doi.org/10.1111/j.1365-246X.1997.tb01215.x

Shi, Y., & Wang, C. Y. (1986). Pore pressure generation in sedimentary basins: Overloading versus aquathermal. *Journal of Geophysics Research: Solid Earth., 1986*(91), 2153–2162.

Snippe, J., Kampman, N., Bisdom, K., Tambach, T., March, R., Maier, C., Phillips, T., Inskip, N. F., Doster, F., & Busch, A. (2022). Modelling of long-term along-fault flow of CO_2 from a natural reservoir. *International Journal of Greenhouse Gas Control, 118*, 103666.

Vannucchi, P. (2020). Scaly fabric and slip within fault zones. *Geosphere, 15*(2), 342–356.

Wang, C., Elsworth, D., Fang, Y., & Zhang, F. (2020). Influence of fracture roughness on shear strength, slip stability and permeability: A mechanistic analysis by three-dimensional rock modeling. *Journal of Rock Mechanics and Geotechnical Engineering, 12*, 720–731.

Wibberley, C. A. J., & Shimamoto, T. (2003). Internal structure and permeability of major strike-slip fault zones: The Median Tectonic Line in Mie Prefecture, Southwest Japan. *Journal of Structural Geology, 25*, 59–78. https://doi.org/10.1016/s0191-8141(02)00014-7

Witherspoon, P. A., Wang, J. S. Y., Iwai, K., & Gale, J. E. (1980). Validity of cubic law for fluid flow in a deformable rock fracture. *Water Resources Research, 16*(6), 1016–1024.

Xu, S., Fukuyama, E., Yamashita, F., Kawakata, H., Mizoguchi, K., & Takizawa, S. (2023). Fault strength and rupture process controlled by fault surface topography. *Nature Geoscience, 16*(1), 94–100. https://doi.org/10.1038/s41561-022-01093-z

Xue, L., Li, H. B., Brodsky, E. E., Xu, Z. Q., Kano, Y., Wang, H., Mori, J. J., Si, J. L., Pei, J. L., Zhang, W., Yang, G., Sun, Z. M., & Huang, Y. (2013). Continuous permeability measurements record healing inside the Wenchuan earthquake fault zone. *Science, 340*, 1555–1559. https://doi.org/10.1126/science.1237237

Yang, Y., & Dunham, E. M. (2021). Effect of porosity and permeability evolution on injection-induced aseismic slip. *Journal of Geophysical Research Solid Earth, 126*, e2020JB021258. https://doi.org/10.1029/2020JB021258

Ye, Z., & Ghassemi, A. (2018). Injection-induced shear slip and permeability enhancement in granite fractures. *Journal of Geophysical Research: Solid Earth, 123*(10), 9009–9032. https://doi.org/10.1029/2018JB016045

Ye, Z., & Ghassemi, A. (2019). Injection-induced propagation and coalescence of preexisting fractures in granite under triaxial stress. *Journal of Geophysical Research: Solid Earth, 124*(8), 7806–7821. https://doi.org/10.1029/2019jb017400

Yielding, G., Freeman, B., & Needham, D. T. (1997). Quantitative fault seal prediction. *American Association of Petroleum Geologists Bulletin, 81*(6), 897–917. https://doi.org/10.1306/522B498D-1727-11D7-8645000102C1865D

Yu, H., Zhang, W., Zhang, Z., Li, Z., & Chen, X. (2020). Investigation on the dynamic rupture of the 1970 M7.7 Tonghai, Yunnan, China, earthquake on the Qujiang Fault. *Bulletin of the Seismological Society of America, 110*(2), 898–919. https://doi.org/10.1785/0120190185

Zhang, S., Tullis, T. E., & Scruggs, V. J. (1999). Permeability anisotropy and pressure dependency of permeability in experimentally sheared gouge materials. *Journal of Structural Geology, 21*(7), 795–806.

Zheng, J., Zheng, L., Liu, H. H., & Ju, Y. (2015). Relationships between permeability, porosity and effective stress for low-permeability sedimentary rock. *International Journal of Rock Mechanics & Mining Sciences, 78*, 304–318.

Zhu, W., & Wong, T.-F. (1997). The transition from brittle faulting to cataclastic flow: Permeability evolution. *Journal of Geophysical Research, 102*, N0. B2, 3027–3041.

Zimmerman, R., & Bodvarsson, G. (1996). Hydraulic conductivity of rock fractures. *Transport in Porous Media, 23*, 1–30.

Chapter 4
Fault Hydromechanical Behavior and Induced Seismicity

Abstract In this chapter, we explore how fault hydromechanical behavior can impact seismic rupture. Our focus is to describe how an initially inactive fault hidden in a basin can be activated by CO_2 injections. Moreover, we explore if there is any parameter specific to CO_2 storage that has to be considered in induced seismicity. In Sect. 4.1, we first describe the most current framework used to define fault plastic instability which is based on the rate-and-state friction model. Second, we describe an alternative concept based on "standard" plasticity (Piau et al. in J Geophys Res Solid Earth, 125, 2020). We discuss how this alternative model may me more general than the rate-and-state friction model. In Sect. 4.2, we describe how the rate-and-state friction physics has been implemented into fluid injection numerical models and applied to describe induced seismicity at basin scales. Section 4.3 is an attempt to define some specificities of seismicity induced by CO_2 storage. In Sect. 4.4, we test the effects of some CO_2 injection specificities using the "standard" plasticity model from Piau et al. (J Geophys Res Solid Earth, 125, 2020). We highlight the possible strong influence of the brine compressibility evolution with CO_2 saturation on fault seismic rupture.

Keywords Seismic rupture · Fluid injections · Rate-and-state friction · Plasticity · Background stress rate · Fault permeability · Fluid properties · Basin scale scenarios

4.1 Transition from Stable Fault Rupture to Unstable Seismic Rupture

Transition from stable fault rupture to unstable seismic rupture can be defined when the fault strength decreases with increase in slip velocity and ends up into a self-driven slip acceleration. This means that earthquakes initiate after a nucleation period of fault strength decrease that may take seconds to years. Dieterich (1994) considers that the nucleation period mainly depends on the initial slip over the nucleation source area and on the stressing history of the fault that depends on the nucleation

© The Author(s), under exclusive license to Springer Nature Switzerland AG 2025   57
Y. Guglielmi, *A Review of CO2 Storage Integrity and Fault Zone Risk*,
SpringerBriefs in Earth System Sciences, https://doi.org/10.1007/978-3-031-81529-4_4

source internal processes and on external processes, like stress modifications caused by other surrounding earthquakes, aseismic slip or fluid pressurization. At seismic failure, fault slip is almost not dependent on fault strength evolution. It is controlled by inertia forces that mainly dissipate as seismic waves and thermal heat. Here, we compare two frameworks used to define seismic instability of faults (Fig. 4.1), respectively the rate-and-state friction model that considers a fault as a zero-thickness frictional interface (RST, Dieterich, 1994) and a "standard" plasticity model that considers the fault as a plastic zone with a pre-existing thickness (Piau et al., 2020).

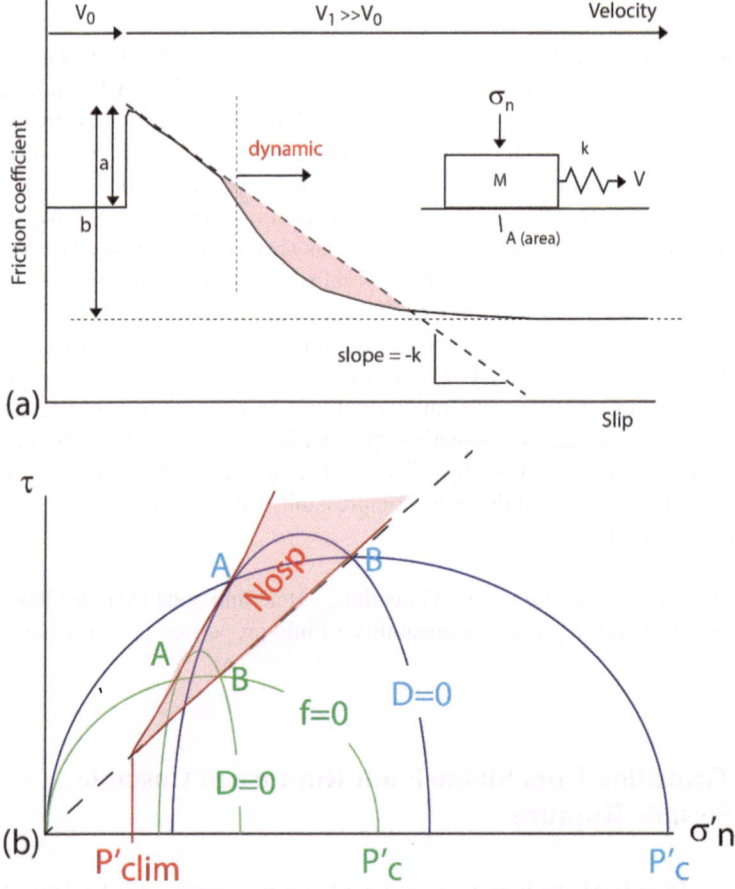

Fig. 4.1 Transition from stable to dynamic fault rupture (red area) described with **a** the rate-and-state friction physics and **b** the Cam-Clay plastic instability physics. **a** In the rate-and-state physics, a dynamic instability occurs when the frictional strength drops rapidly below the stiffness k of the spring-slider system. **b** In Cam-Clay physics, the no-static plastic evolution triangular area is delimited from solutions of the ($f = 0$, $D = 0$) system of equations (from Piau et al., 2020). When stress path reaches the area delimited by the thick red line, rupture is self-driven by inertial forces, potentially corresponding to a seismic rupture

- Seismic rupture defined from the rate-and-state frictional physics

The most common framework to describe the evolution of a frictional interface to unstable seismic (i.e., dynamic) rupture is the rate-and-state friction theory (RST, Dieterich, 1978). RST is deduced from laboratory direct shear experiments that can schematically be represented as a spring-slider system (Fig. 4.1a). In these experiments, a rigid solid with a mass (M) is pulled at a velocity (V) on a horizontal interface. A stress is applied on the solid, perpendicular to the interface and kept constant during the test (normal stress σ). A spring figures the elastic shear stiffness (K) of the system that is the ratio between the shear modulus of the block and the length of the interface (A). The elastic stiffness represents the elastic stiffness of the crust around the fault. The solid of mass (M) per unit area is initially pulled on the interface at a background velocity (V_0). A step increase in velocity from V_0 to V_1 is suddenly applied to the solid and kept constant until the end of the experiment.

The frictional coefficient (μ) of the interface varies with the slip rate (V_1) and with the evolution of the state of the interface asperities (θ) with time. This is commonly described by Eqs. 4.1 to 4.3:

$$\mu(V, \theta) = \mu_0 + a \ln\left(\frac{V}{V_0}\right) + b \ln\left(\frac{V_0 * \theta}{D_c}\right), \tag{4.1}$$

where μ_0 is the friction coefficient measured at displacement speed V_0. The parameters (a) and (b) respectively describe the instantaneous friction increase at the velocity step (direct effect) and the nonlinear transient friction decrease at constant velocity V_1 (state effect). D_c is the slip distance over which the friction evolves to a new steady state. There are two commonly used laws to describe the asperity aging state parameter θ, respectively the aging law (Dieterich, 1979)

$$\frac{d\theta}{dt} = 1 - \frac{V * \theta}{D_c}, \tag{4.2}$$

and the slip law (Ruina, 1983).

$$\frac{d\theta}{dt} = -\frac{V\theta}{D_c} \ln\left(\frac{V\theta}{D_c}\right) \tag{4.3}$$

These laws give about the same θ evolution when friction is close to steady state. At steady state, $\theta = \frac{D_c}{V}$ and the friction equation resumes to

$$\mu(V) = \mu_0 + (a - b) \ln\left(\frac{V}{V_0}\right). \tag{4.4}$$

This means that the friction dependency on the slip rate is described by $(a - b)$. If $(a - b) > 0$, the friction is increasing with slip rate. This behavior is called "velocity-strengthening" and leads to stable, aseismic fault slip. If $(a - b) < 0$, the friction is decreasing with slip rate. It is called "velocity-weakening" and leads to potential frictional instability with accelerating slip to earthquake. Laboratory values for $(a - b)$ and Dc lie in the range of $\pm 10^{-3}$ to 10^{-2} and 100-to-300 µm. $(a - b)$ values and sign depend on many different experimental conditions such as temperature, strain rate, imposed normal stress and pore fluid properties, and the mineralogy composition. RST physics has been extensively used to describe a large variety of fault rupture characteristics and behavior from slow, aseismic slip to dynamic, seismic rupture (see some examples in Blanpied et al., 1998; Collettini et al., 2011; Ito & Ikari, 2015; Marone, 1998; Niemeijer et al., 2016).

From a theoretical point of view, with the RST physics, slow sliding on the fault occurs before instability. Two "mandatory" conditions are needed for frictional instability to lead to seismic rupture, corresponding to the transition from quasi-static to dynamic slip:

- $(a - b) < 0$ to favor frictional instability potentially leading to earthquake (while $(a - b) > 0$ favors frictional stability and no seismicity, Ruina, 1983).
- The interface elastic stiffness K must be lower than a critical stiffness K_c (Rice & Ruina, 1983; Ruina, 1983).

Figuring the fault as a simple mechanical spring-block model, the transition from quasi-static to dynamic slip can be figured as if a sudden stress perturbation $q(t)$ occurring at $t = 0$ was applied to the sliding block (Fig. 4.1a). It can be illustrated as a sudden shortening $x(t)$ of the spring from its steady state length. Rice and Ruina (1983) use the equation of motion (4.5) to describe the block acceleration as a function of the spring stiffness K, shear stress τ and the stress perturbation $q(t)$.

$$M \frac{d^2x(t)}{dt^2} = \tau_0 - kx(t) - \tau(t) + q(t), \tag{4.5}$$

where $x(t)$ is the change in the slider position and $\frac{d^2x(t)}{dt^2}$ is the acceleration of the slider caused by the $q(t)$ stress perturbation. τ_0 is the steady state value of shear stress before imposing stress perturbation. Equation (4.5) can be linearized assuming that $q(t)$ has the following form $q(t) = q_0$ heaviside(t) that returns q_0 for $t > 0$ and 0 for $t < 0$. At $t > 0$, the fault slip velocity is $V(t) = V_0 + \frac{dx(t)}{dt}$, while the state variable varies of $\theta(t) = \frac{D_c}{V_0} + \xi(t)$ (Roy & Marone, 1996). These equations can be substituted in (4.4), in addition to considering RST Eqs. (4.3) and (4.4) that describe the friction coefficient evolution with velocity.

$$M \frac{d^2x(t)}{dt^2} = \tau_0 - kx(t) - \tau(t) + q(t) = \mu_0\sigma - kx(t) - \mu\sigma + q_0$$

$$= \mu_0\sigma - kx(t) - \left[\mu_0 + a\ln\left(\frac{V(t)}{V_0}\right) + b\ln\left(\frac{V_0\theta}{D_c}\right)\right]\sigma + q_0$$

$$= -kx(t) - a\sigma\ln\left(1 + \frac{\frac{dx(t)}{dt}}{V_0}\right) - b\sigma\ln\left(1 + \frac{V_0\xi(t)}{D_c}\right) + q_0 \tag{4.6}$$

If we then use the Taylor series and neglect all terms that are higher than second order, we get the final Eqs. (4.7) and (4.8).

$$M \frac{d^2x(t)}{dt^2} = \sigma \left[-\frac{k}{\sigma} x(t) - \frac{a \frac{dx(t)}{dt}}{V_0} - \frac{bV_0 \xi(t)}{D_c} \right] + q_0 \qquad (4.7)$$

and

$$\frac{d\theta(t)}{dt} = 1 - \frac{V(t)\theta}{D_c} = 1 - \frac{\left(V_0 + \frac{dx(t)}{dt}\right)\left(\frac{D_c}{V_0} + \xi(t)\right)}{D_c} \approx - \left[\frac{V_0 \xi(t)}{D_c} + \frac{\frac{dx(t)}{dt}}{V_0} \right].$$

$$(4.8)$$

The quasi-static motion is defined in Eq. (4.7) when the acceleration term $\frac{d^2x(t)}{dt^2} \approx$ 0. In that case, fault motion $x(t)$ depends on the frictional terms $-\frac{a \frac{dx(t)}{dt}}{V_0} - \frac{bV_0 \xi(t)}{D_c}$. The dynamic motion is when fault movement is dominated by acceleration (inertial term) while friction terms in Eq. (4.7) become negligible. Equation (4.7) then approximates to the one of a harmonic oscillator (4.9)

$$M \frac{d^2x(t)}{dt^2} = -kx(t), \text{ where } x(t) = x_0 \exp(i\omega t) \text{ with } \omega = \sqrt{\frac{k}{M}}. \qquad (4.9)$$

From a linear stability analysis of Eq. (4.7), Rice and Ruina (1983) showed that the transition from quasi-static to dynamic motion depends on the system's critical stiffness (Eq. 4.10).

$$K_c = \frac{\sigma(b-a)}{D_c} \left[1 + \frac{MV_0^2}{\sigma a D_c} \right] \qquad (4.10)$$

Roy and Marone (1996) highlight that when in quasi-static motion, the velocity is slow and the term $\frac{MV_0^2}{\sigma a D_c} \ll 1$ can be neglected in Eq. (4.10). This allows defining an upper-bound velocity for quasi-static slip from the $\frac{MV_0^2}{\sigma a D_c}$ term of Eq. (4.10), by considering that when this term gets close to 1, it is the onset of dynamic slip.

$$V_{qs} = \sqrt{\frac{\sigma a D_c}{M}} \qquad (4.11)$$

In addition, by setting the friction terms $-\frac{a \frac{dx(t)}{dt}}{V_0} - \frac{bV_0 \xi(t)}{D_c} = 0$ and combining with Eq. (4.9), Roy and Marone (1996) define a limiting velocity at which slip becomes inertia-dominated.

$$V_{in} = \frac{a D_c}{(b-a)} \sqrt{\frac{k}{M}} \qquad (4.12)$$

Thus, Roy and Marone (1996) define three slip velocity regimes during a seismic nucleation model:

- Quasi-static slow slip velocity regime ($V < V_{qs}$) where there is negligible acceleration above the background velocity.
- Quasi-dynamic regime ($V_{qs} < V < V_{in}$) where slip velocity starts to significantly increase above the background velocity while friction decays due to velocity.
- Dynamic "seismic" slip regime ($V > V_{in}$) characterized by a high slip acceleration that occurs when the velocity weakening causes friction to fall below the applied stress (below the dashed line of slope-k in Fig. 4.1a). Thus, if the fault has rate-weakening properties and the reduction of shear strength is fast, an earthquake can occur once the slipping region reaches a critical size.

Roy and Marone's model is sensitive to the following parameters which values are considered to correspond to faults at seismogenic depths while friction parameters were measured in the laboratory, $V_0 = 10^{-9}$m/s, $\sigma = 100$MPa, $k = 1$-to-5 GPa/m, $M = 10$ to 10^7 kg/m^2, $a = 0.005$ to 0.007, $b = 0.006$ to 0.008 and $D_c = 2 \times 10^{-5}$ to 2×10^{-3}m. With these parameters, the two instability criteria are satisfied with $k/k_c = 0.2$ to 0.9 and a-$b < 0$, respectively. For the same perturbation q_0, a larger M causes a smaller pre-seismic slip and a longer pre-seismic duration. A higher stiffness, a longer Dc and a larger a yield a larger pre-seismic slip and a longer pre-seismic duration. Dynamic slip regime is estimated to occur at V_{in} of 0.009 to 0.8 m/s depending on the parameters cited above. More recently, Im et al. (2019) confirmed these sensitivity study results as well as the validity of Eq. (4.8) to pre-figure the system instability.

Likewise, Van den Ende et al. (2018) compared these empirical rate-and-state friction models with microphysical numerical models. Although their work concerned fault slip at large seismogenic depths and very high slip velocities, it highlighted the importance of transitions between slip velocity regimes on nucleation and arrest of rupture. Indeed, they showed possible transitions between velocity weakening and strengthening due to switches in mechanical processes, (i.e., from pressure-solution ending in strengthening at relatively low slip velocities to frictional heating ending in weakening at higher velocities). They pointed out that such transitions that could explain seismic slip arrest or aseismic slip above V_{in} required a variation of rate-and-state parameters that needed to be implemented as constitutive laws inside advanced numerical models. This is usually neglected in most induced seismicity studies. Moreover, some mechanisms are rarely considered in RST-based approach of seismic rupture, such as the strengthening role of dilatancy (Segall et al., 2015), the weakening role of creep-induced compaction (Yang & Dunham, 2023), the weakening role of the decrease in fault pore fluid compressibility (Cornelio et al., 2020; Sleep & Blanpied, 1994) and the transient change of rate-and-state parameters with the amplitude and rate of fluid pressurization (Cappa et al., 2019). In Sect. 4.3 we will review some of these factors in more details, selecting the ones more suitable to influence seismic rupture at relatively moderate depths and in initially relatively low tectonically active areas.

- Seismicity defined from plastic instability.

Piau et al. (2020) developed a fault zone conceptual model defined as a poro-plastic Cam-Clay interface surrounded by two symmetrical elastic pads to investigate different scenarios of fault quasi-static to dynamic rupture (Figs. 2.2 and 4.1b). For details on Piau et al. (2020) model physics see Chap. 2. Compared to the RST physics that is mainly focused on the velocity dependency of the fault friction, this constitutive model is more general since it accounts for the combined effects of the full in situ stress tensor, tectonic loading, contracting/dilatant deformation and fluid conditions (drained/undrained, compressibility). Here we focus on how dynamic rupture is defined based on model's plastic instability which is when rupture is self-driven by inertial forces. Two conditions are required:

- Failure occurs in the dilatant domain which corresponds to the plastic flow law $f = 0$;
- The time derivative of the plastic multiplier $\dot{K}$ becomes negative.

At failure $f = 0$, the plastic multiplier relates stress variation to plastic displacement of the upper fault compartment (upper pad), as formulated by Piau et al. (2020) in Eqs. (4.13)–(4.15).

$$D\dot{K} = N, \tag{4.13}$$

with N being the calculated plastic displacements given by (4.14)

$$N = (2\sigma_{yy\prime} + P_{c\prime})\left(R(u_{ext} - u) + \frac{\lambda + 2\mu}{h}\dot{V}_2\right) + \frac{6\mu}{M^2 h}\sigma_{xy}\dot{U}_2, \tag{4.14}$$

and D being the stresses at plastic failure given by (4.15)

$$D(\sigma_{yy\prime}, \sigma_{xy}, S, p_{c\prime})$$
$$= \left(\frac{\lambda + 2\mu}{h} + S\right)(2\sigma_{yy\prime} + p_{c\prime})^2 + \frac{36\mu}{M^4 h}\sigma_{xy}^2 + \alpha p_{c\prime}(2\sigma_{yy\prime} + p_{c\prime})\sigma_{yy\prime}. \tag{4.15}$$

For a quasi-static plastic solution, the inequality $\dot{K} \geq 0$ must be verified. If we consider that fault is driven to instability by a far-field shear displacement $\dot{U}_2$, it means that $\dot{U}_2$ and σ_{xy} have the same sign. If they had opposite sign, it would lead to a progressive unloading driving the fault to stability. Then, with $N \geq 0$ the condition for $\dot{K} \geq 0$ requires $D > 0$. Thus, Piau et al. (2020) define plastic instability when $f = 0$ and $D < 0$. They then solve the system of two equations $f = 0$ and $D = 0$ for different consolidation pressures $p_{c\prime}$. This gives a series of solutions $(p_{c\prime}, A, B)$ that allow defining a triangular domain in the Mohr diagram that they call the Ω_{Nosp} (Nosp = No static plastic evolution, Fig. 4.1b).

Figure 4.1b shows that it is possible to define the smallest value of consolidation pressure for which $f = 0$ and $D = 0$, and the fault is likely to exhibit a dynamic plastic unstable evolution (Eq. 4.16).

$$p'_{clim} = \frac{6\mu}{\alpha h M^2} \left(1 + \sqrt{\frac{2(\lambda + 2\mu + Sh)}{3\mu}} \right), \tag{4.16}$$

where h is the fault zone thickness, S is the coefficient relating fault pore pressure to dilation, $M = \sqrt{3} \tan \emptyset_c$ and (λ, μ) are Lame's coefficients describing poroelastic coupling into the pads.

Piau et al. (2020) then developed a solution considering that the fault dynamic displacement is described by the pad vibrations at each space (x) and time (t) under material properties (c, ρ). Tangential and normal stresses to the fault interface are related to vibrations using the elastic lame parameters of the pad. This allows calculating seismic slip, normal displacement, tangential and effective normal stress. The duration of the dynamic phase depends on the number of reflections of P- and S-waves at the pad's boundaries, or in other words on how kinetic energy is dissipated into the pads (which is not considered in this plastic dynamic model). Nevertheless, Piau et al. (2020) give different solutions, including a possibility to estimate the stresses and fault displacements at the end of the dynamic phase, and the possibility to consider infinite pads.

4.2   Modeling Seismicity Induced by Fluid Injections

To account for seismicity induced by fluid injections, most of modeling approaches first consider the effects of pore pressure and poroelastic stress on changes in Coulomb failure stress and then use the Dieterich's rate-and-state nucleation model to assess for the temporal evolution of the induced seismicity rate (Segall & Lu, 2015). A forward poroelastic (Chang & Segall, 2016) or fully coupled hydromechanical model (TOUGH-FLAC3D, Luu et al., 2022) is, for example, used to calculate the spatio-temporal Coulomb stress changes (ΔCFS) induced by fluid injections. The stress rate is approximated from the calculation of the Coulomb stress change versus time since the beginning of the injection. The forward model has fixed displacement basal boundary conditions while a stress gradient versus depth is applied on model vertical faces. Before injection, this model is run to mechanical equilibrium. Stress variations are then calculated by loading the equilibrated model with a fluid injection time history (for example by imposing an injection flowrate inside or on one side of the model).

$$\Delta \text{CFS} = \Delta \tau_s + \mu (\Delta \sigma_n + \Delta P), \tag{4.17}$$

where (τ_s, σ_n, P) are the shear and normal stresses and pore pressure calculated on a given fault plane. μ is the friction coefficient.

The Coulomb stressing rate is defined as

$$\dot{t} = \frac{\text{d}\Delta \text{CFS}}{\text{d}t}. \tag{4.18}$$

Then, the calculated stressing rate is introduced into a second model that considers Dieterich's rate-and-state nucleation physics and that allows calculating the seismicity rate $\frac{dR}{dt}$. As shown in Eq. (4.19), this model is intrinsically unstable since it describes the change from a natural seismicity rate to a seismicity rate induced by fluid injections. Thus, this type of model postulates that the crust during the pre-injection period contains a population of seismic sources that on average produce a constant seismicity rate under a constant background tectonic stressing rate. Dieterich (1994) assumes that each of these sources correspond to a quasi-dynamic regime where fault slip velocity starts to significantly increase above the background velocity while friction decays due to slip velocity (see previous Sect. 4.1). Segall and Lu (2015) suggest the following simplified Eq. (4.19) for the seismicity rate (modified from Dieterich, 1994).

$$\frac{dR}{dt} = \frac{R}{t_c}\left(\frac{\dot{t}}{\dot{t}_0} - R\right), \tag{4.19}$$

where R is the ratio of the observed seismicity rate R_d relative to the background seismicity rate r_0 at the $\dot{t}_0$ background stress rate, i.e., the stress rate related to natural tectonic activity before the start of the anthropogenic fluid injections. Time t_c is the characteristic time decay. It can be defined as

$$t_c = \frac{a\sigma_n'}{\dot{t}_0}, \tag{4.20}$$

where a is a rate-and-state parameter that describes the instantaneous friction increase associated to a slip velocity step (see Eq. 1.1). $\sigma_{n'} = \sigma_n - P$ is the effective normal stress.

This adaptation of the rate-and-state physics to relate the Coulomb stress change on natural faults to changes in seismicity rates has some intrinsic limitations:

- As explained in Sect. 4.1, the RST physics describes the transition from the aseismic slip nucleation phase to the potentially seismic slip acceleration. It thus "predicts" the potential occurrence of the earthquake mainshock, but it does not consider complex interactions between earthquakes during the aftershock sequences that thus are not predicted. This implies that the induced seismicity field catalogue must be de-clustered for the approach to be applied only on the main events. In addition, this approach can eventually match the observed main events seismicity rate, but it does not give an estimate of the events magnitude. However, combined geomechanical and seismicity rate models allow estimates of earthquake frequency occurring within a given stimulated volume. A magnitude-frequency relationship can then be used to estimate magnitude with some additional assumptions that such relationships remain homogeneous in both time and space (Zhai et al., 2019).
- As described above, two models are combined but not coupled to estimate the induced seismicity rate, respectively a Coulomb failure (CF) quasi-static model

followed by a rate-and-state (RST) seismicity model. The CF model is in initial mechanical equilibrium before the injection starts. It thus means that it does not consider the effect of the background stress rate $\dot{\tau}_0$ on fault stability evolution before injection. The RST model is quasi-dynamic since it considers that fault friction evolves with time and stressing history. In this model, the fault is already at failure and injection will "just" eventually accelerate failure. Indeed, Eq. (4.18) assumes that the Coulomb stressing rate is independent of the background, pre-injection stress state (Segall & Lu, 2015). The main assumption is that the increase in the stressing rate induced by injections is much larger than the effect of that the low background stressing rate may have on the pre-injection fault state.

4.3   Key Factors to Consider in Seismicity Induced by CO_2 Storage

Equations in Sects. 4.1 and 4.2 highlight the key factors considered in models to predict seismicity induced by fluid injections (i.e., the background stress and seismicity history (Eqs. 4.19 and 4.20) that include both the eventual tectonic activity and previous anthropogenic exploitation, the pore pressure build up and the associated poroelastic stress). Here we discuss these factors in the context of CO_2 storage.

• Low background stress rate ($\dot{\tau}_0$) and low background seismicity rates (r)

The background stress and seismicity history of the upper crust are cornerstone parameters of the seismicity rate theory since this theory is based on estimating the change in a background seismicity rate induced by fluid injections (Dieterich, 1994; Segall & Lu, 2015). In practice, ($\dot{\tau}_0$, r_0, $\sigma_{n'}$, a) from Eqs. 4.19 and 4.20 are considered as initial properties. $\sigma_{n'}$ is considered constant and it is deduced from a geomechanical model stress calculation before injection loading (note that some authors include the possibility for $\sigma_{n'}$ to vary during the injection period in the seismicity rate estimation). The seismicity rate R_d is then calculated from the Coulomb stress rate $\dot{\tau}$ using Eq. 4.19. Then, ($\dot{\tau}_0$, r_0, $\sigma_{n'}$, a) are "adjusted" using optimization techniques in order to best fit the modeled seismicity rate R_d to the measured one.

The problem is that for fluid storage or extraction in low active tectonics intraplate basins, background stress and seismicity rates are low (and hard to estimate). For example, in the cases of the Groningen gas field in Netherlands (Dost et al., 2017), the Illinois Basin (Galgana & Hamburger, 2010) or in Oklahoma (Ellsworth, 2013) there was almost no detected seismicity prior to injections. The 2×10^{-6} and 20×10^{-3} MPa/yr background stress rates estimated in these regions may be not sufficient to trigger a steady state background seismicity rate as defined in Dieterich (1994).

In Groningen, Candela et al. (2019) considered that the background seismicity was at steady state after the first 25 years of reservoir exploitation from 1968 to 1993, to analyze the increase in the seismicity rate during the following 23 years from 1993 to 2016. After optimization of ($\dot{\tau}_0$, r_0, a) to best match the model and measured R_d variation during the 1993–2016 period they get values respectively of (0.0015 to 0.02 MPa/yr, 0.075 to 0.4 events/yr, 0.14 to 0.8). First, the values for parameter "a" are high compared to values of 0.001 to 0.003 usually measured in the laboratory. Second, they get large nucleation times t_c of 87–6700 years when the above ($\dot{t}_0$, a) values are used in Eq. 4.19, for an average effective normal stress of 12.5 MPa on the faults affecting the reservoir. In other words, they highlight that in 1993 the Groningen reservoir's faults could not be at a steady state seismicity rate because the nucleation times are much larger than the duration of 1968–1993 reservoir exploitation period.

The Illinois case study allows comparing two approaches to "constrain" the background stress and seismicity rate to analyze the current Galgana and Hamburger (2010) make the hypothesis that the diffuse seismicity observed in Illinois results from a 190-year long aftershock sequence following the 1811–1812 New Madrid earthquakes. They develop a geodynamical basin scale model that contains two 40 and 60 km long vertical faults on which 4 and 5m slip are suddenly triggered along fault strike to approximately figure the 1811–1812 Mw 7.2–8.2 earthquakes sequence. A 25km thick elastic lithosphere is assumed overlying a viscous asthenosphere with a 10^{-19} Pa s viscosity. They calculate the stress rate following the earthquake and convert it to seismicity rate using the Dieterich (1994) formulation. To best fit the calculated and observed seismicity rates they vary the asthenosphere viscosity, the fault frictional property a and the effective normal stress $\sigma_n\prime$. After 190 years of calculated stress relaxation, the best fit results in a background stressing rate of 5 Pa/yr corresponding to a $5 \cdot 10^{-9}$ strain rate (as measured by surface GPS), a low effective normal stress of 5 MPa and an a $= 0.01$.

In comparison, Luu et al. (2022) conduct a modeling protocol like the ones used by Candela et al. (2019). In Illinois, they use TOUGH-FLAC multiphase fluid flow and geomechanical simulator with ~ 16 vertical fault zones that are 10m thick and a few kilometers long. Fault zones are located as indicated by the observed seismicity alignments. The model contains multiples sedimentary caprock-reservoir layers overlying the basement. Faults are connected to the CO_2 storage reservoir. Faults and layers are given hydromechanical properties according to data available on the site. The model is mechanically consolidated until equilibrium by applying a vertical stress gradient while a no-displacement boundary is given to the model bottom. Then, a supercritical CO_2 injection was applied with the same characteristics as the two injections were conducted in the field from November 2011 to November 2014 and from April 2017 to April 2018. The TOUGH-FLAC simulator allows for coupled hydromechanical simulation of pore pressure, shear and effective normal stress changes with injection time. The model can thus integrate both the effects of pore pressure and poroelastic stresses on the Coulomb stress rate averaged at each fault zone. Equation 4.19 is used to calculate the seismicity rate from the calculated Coulomb stress rate. This is done through a second model written in Python. The authors use the background stress rate of 5 Pa/year from Galgana and Hamburger (2010). Because there are only 8 earthquake recorded over the 18months preceding the first CO_2 injection, the authors use the seismicity induced by the first injection instead, thus making the strong hypothesis that background seismicity was at steady state at the end of the first injection.

With about the same fitting protocol as Candela et al. (2019) they get a background seismic rate r_0 of 0.35 events/yr, an effective normal stress of 2.95 MPa and a value of 0.01. Estimates of background stress and seismicity rates from Hamburger et al. (2010) and from Luu et al. (2022) give nucleation times t_c respectively of 10,000 and 5,900 years. In addition, they rely on effective normal stress values for deep faults that are anomalously low and explain that it might be related to fluid overpressures (although there is no indication about this in the literature). Overall, the high t_c values show that seismicity could not be at steady state when the second injection period started, for example.

These two detailed examples highlight the limitations of the Dieterich (1994) model to describe changes in seismic rates caused by fluid injections in intraplate setting. Note that the Dieterich framework may work better in more tectonically active regions (Jia et al., 2020). In low active tectonics regions, the faults may be initially far from failure, and progressively brought to failure by extraction induced stress. Heimisson et al. (2022) suggest a new Coulomb threshold rate-and-state model for fault activation. Compared to the Dieterich, 1994 model, this model does not consider any background seismicity rate. The faults are initially stable and progressively brought to seismic instability by stress variation induced by injections. Applied to the Groningen case, this type of model apparently better fits the observed time-varying seismicity rate and reproduces better the onset, peak and decline of the observed seismicity rate. In a different way, Zhai et al. (2019) define a time to charge the fault system before which stressing rate can relate to the seismicity rate (which is also a way to define a threshold).

Nevertheless, how initially stable faults can progressively be brought to seismic instability remains poorly understood. It does not seem that poroelastic stress transfer might be a sufficient mechanism as calculated by Luu et al. (2022) and Zhai et al. (2019) even if combined with pore pressure increase it may significantly change the seismicity rate (Segall & Lu, 2015). An alternative mechanism could be the effect of aseismic slip in significantly redistributing stresses in regions way beyond the pressurized regions (Cebry et al., 2022). Indeed, observations on field scale experiments showed that during fault activation by fluid injection, a lot of aseismic slip is produced compared to seismicity in regions way beyond the pressurized areas (Cappa et al., 2019; Guglielmi et al., 2015). Aseismic slip can be produced after Coulomb failure and slip can eventually accelerate to dynamic failure as described in Sect. 4.1. Unfortunately, there is only a few evidence of the role of aseismic slip on faults in deep basins environments (Das & Zoback, 2013). It may also happen that past fluid extraction history initiated these aseismic stress redistributions, preparing the system for seismic instability. One key mechanism could there be differential compaction accommodated by aseismic fault movements for example (Jeanne et al., 2020; Bourne and Oates, 2017).

- Reservoir and fault permeability control

In an exhaustive review of seismicity induced by CO_2 storage, Cheng et al. (2023) show that most of the current theories highlight the key role of pore pressure increase induced by the storage at the basin scale. In the existing CO_2 storage case studies, relatively small amount of induced seismicity of moderate $\ll 3$ magnitudes, is explained by the high permeability of storage reservoirs that favor pressure diffusion rather than pressure build up. Indeed, taking the example of the Illinois Basin, Zhou et al. (2010) modeled a moderate pressure build up of 3.5 MPa caused by 20 injection sites corresponding to a total annual injection rate of 100 Mt CO_2 over 50 years. Nevertheless, they also showed that pressure perturbations propagate quickly away from injection boreholes and that the different injection sites start interfering with each other less than a year after the injection starts. In Oklahoma, Keranen et al. (2014) calculated a ~ 4 MPa maximum pore pressure increase in the Arbukle group where ~ 4 million barrels/month of waste waters were injected between 1995 and 2012. Calculated pressure perturbation extended ~ 35km away from the injection points into the Arbukle group and into the underlying upper basement where most of the induced seismicity was located. A ~ 0.07 MPa pressure threshold above which earthquakes are triggered was deduced from calculating the pore pressure change at the location of the observed earthquakes hypocenters. They calculated that pore pressure propagated over about 10km in 3 years, considering a very high crustal hydraulic diffusivity of 1-to-4 m^2/s, which is in the range of values estimated from the inversion of seismic swarm migration data (Shapiro, 2015).

Since most of the seismicity occurs in the basement below the injection sites, the question of the basal diffusivity appears to be crucial, and values of 1 to 4 m^2/s that are deduced from indirect seismological approaches look high compared to what is estimated from other types of methods. Indeed, Zhang et al. (2013) considered an average permeability of 2 10^{-17} m^2 and a specific storage of 10^{-7} /m in a large-scale study of the effects of the hydrogeology of the mid-continent basement of the USA on induced seismicity. This corresponds to a $\sim 10^{-3}$ m^2/s diffusivity, way below the ones previously cited. They also showed that in the case of the basement directly hydraulically connected to the injection reservoir, this permeability could favor pore pressure diffusion and brittle failure in the upper basement limit. In this hydrogeological context, a fault zone connected to the reservoir and rooted in the basement that would have 5 orders of permeability higher than basement (i.e., 10^{-13} versus 10^{-17} m^2) would increase the depth of basement brittle failure by about a factor of 6. Guo et al. (2021) recently used earth tides to estimate a 2 to 7 10^{-14} m^2 permeability of a deep basement fault in Nevada, confirming that such high-permeable fault zones could affect the deep crystalline crust. Brixel et al. (2020) found similar fault zone permeability values from direct hydraulic pulse tests conducted on granite at 1.5km depth in Switzerland.

In a theoretical study, Chang and Segall (2016) used the range of fault zone permeability values from Zhang et al. (2013) to explore the link between fault permeability and the induced seismicity rate. They showed that there may exist two main cases where fault permeability plays a crucial role in propagating pore pressure and triggering increase in seismicity rate:

- High seismicity rate caused by the direct pore pressure diffusion in high-permeable 10^{-13} m^2 faults hydraulically connected to the storage reservoir and rooted in the underlying basement. This could be the case of the induced seismicity triggered by CO_2 injections at the Decatur project in Illinois as modeled by Luu et al. (2022) and in Oklahoma (Kolawole et al., 2019) for example.

• Moderate seismicity rate caused by indirect poroelastic stressing of isolated low permeable 10^{-21} m^2 basement faults. This could represent a majority of "minor" faults (i.e., too small to extend upward from the basement to the injection reservoir). Indeed, most of induced seismicity field studies highlight much more fault lineaments compared to the usually relatively limited number of majors fault zones that potentially hydraulically connect the injection layer with the deep basement, see the Weiburn CO_2 area for example, Verdon (2016) or more recently the Oklahoma area (Kolawole et al., 2019). Moreover, poroelastic stressing may be favored or inhibited on unconnected basement faults depending on how these faults offset the sedimentary layers directly in contact with basement layers and on the elastic properties of these layers. For example, Zhang et al. (2013) show that a continuous soft layer immediately above the crystalline basement may kill the possibility for pore pressure to build-up in basement faults, and thus may limit induced seismicity. Nevertheless, Bondarenko et al. (2022) shows that if this competent layer is offset by a basement fault, the contrast in the elastic properties between the soft sedimentary compartment and the stiff crystalline one may enhance stress perturbation enough to favor induced seismicity by poroelastic stress transfer.

There is another effect that is how permeability changes during a fault activation may impact induced seismicity. Marguin and Simpson (2023) compared a crustal scale fault hydromechanical model with the same dry model to show that a limited fluid overpressure tends to reduce coseismic slip, stress drop, maximum sliding velocity, rupture velocity and the earthquakes recurrence time. It is understood that a large fault porosity increase induced by dilation with slip velocity will dissipate excess pore pressure, increase the effective normal stress and thus slow down the slip (Segall & Rice, 2015). The associated permeability increase influence in earthquake models is less considered. Marguin and Simpson (2023) show that at rupture slip induced dilation may limit seismic instability. But they also show that the associated sudden permeability increase may favor a relatively fast upward overpressure propagation that can eventually trigger aftershock like seismicity. This type of dynamic effect of fault valving is also described by Zhu et al. (2020). Moreover, field injection experiments (De Barros et al., 2019; Guglielmi et al., 2015) and crustal scale seismic swarms' observations (De Barros et al., 2020) show that such a high-pressure propagation in an increasing permeability fault may trigger a complex sequence of fluid-driven aseismic slip episodes separated by phases of pressure build up, and bursts of seismicity caused by slip induced stress perturbation on fault asperities. Ross et al. (2022) describes a four years long seismic swarm occurring at 5 to 9 km depth along an active fault zone in Southern California. They show that seismicity aligns at different depths along the fault surface highlighting contrasts of permeability in horizontal layers inside the fault zone architecture. The breaching of these internal fault zone layers favoring the spatio-temporal propagation of the swarm.

- Effects of changes in fluid properties

In storage reservoir, CO_2 density and viscosity vary significantly with pore fluid pressure increase (Islam & Carlson, 2012; Vilarrasa et al., 2019). Indeed, Ringrose et al. (2022) highlight the specificity of CO_2 that, below about 800 m depth, is in a dense-phase that has a gas-like viscosity (around 0.06–0.07 centipoise) but a fluid-like density (500–800 kg/m^3). The phase change of CO_2 is very sensitive to the relatively shallow crust pressure and temperature conditions.

Jung (2014) highlights the interplay between CO_2 buoyancy and fault permeability, through an exhaustive TOUGH modeling of two faults zones naturally leaking CO_2 in the Colorado Plateau. Supercritical and gaseous CO_2 tend to ascend buoyantly along the fault zone due to its relatively lower density and viscosity than the surrounding brines. Jung's models show that even low permeability faults cannot completely prevent CO_2 ascent toward surface. Snippe et al. (2022) showed that fault relative permeability to CO_2 obviously depends on fault aperture. At low aperture, the non-wetting CO_2 phase cannot penetrate the fault fractures resulting in trapping of large portions of irreducible water around which CO_2 must flow. When aperture gets larger this effect dissipates and viscous forces begin to dominate. There is less interference between the CO_2 and the original pore water phases. In addition, they show that there is a weak dependence of capillary pressure on water saturation and fracture aperture (in kPa range), and thus capillary pressures in fractures can apparently be easily overcome by a CO_2-brine gravitational pressure head.

In the laboratory, Cornelio et al. (2020) conducted shear stress-controlled experiments of a Westerly granite saw-cut fracture using a rotary shear apparatus and varying the fracture pore fluid viscosity from room humidity to pure water and more viscous fluids. Low viscous water or vapor favor weakening under flash heating mechanisms while high viscous fluids favor weakening under elastic-hydrodynamic lubrication. CO_2 supercritical may be in the first case. A phase change to dissolved and a vapor phase would even decrease viscosity and bring back the fault slip from unstable to stable slow slip.

In the field, Geli et al. (2014) relate seismic precursors observations triggered on an oceanic transform active fault to the increase in supercritical fluids compressibility caused by fault pore pressure decrease induced by mechanical dilation under shear. Using a Cam Clay model to figure the dilation of the fault zone, they calculate that shear induced dilation is causing enough pore pressure decrease for a fluid compressibility increase significant enough to trigger fault slip seismic instability.

4.4 Basin-Scale Scenarios of Induced Seismicity

Here we used the plastic dynamic instability as defined by Piau et al. (2020) to explore some scenarios of seismicity induced by a CO_2 injection source. As in Sect. 2.4, we use the code developed by Piau et al. (2020) to explore some parameters that are relevant to CO_2 storage and that can make a fault seismic. We consider the reference

case with a background 0.006 m/yr tangential displacement rate that is described
in Sect. 2.4 (Fig. 2.7 and Table 4.1). The 0.006 m/yr background displacement rate
generates a 0.003 MPa/yr shear stress increase on the fault. This is in the range
of values considered in low active tectonics intraplate basins (Candela et al., 2019;
Dost et al., 2017; Ellsworth, 2013; Galgana & Hamburger, 2010). A constant pressure
injection is conducted for 22.5 years. Fault rupture occurs at 1.4 years after injection
start (point a in Fig. 4.2a). Then, there is a slow 6.7×10^{-4} m/yr fault creep associated
with a 1.4×10^{-4} m/yr fault opening. The fault rupture slightly accelerates with time,
remaining below the background 6×10^{-3} m/yr displacement rate (Fig. 4.3a and b,
and Table 4.2). After 22.5 years of injection, total slip and normal displacement
respectively are 0.015 and 0.0032 m (Table 4.2).

We then explore how the fluid compressibility, the yield law coefficient α, the fault
zone thickness and the porosity may change the creeping behavior described in the
reference case into an accelerated seismic slip. Note that the way these "intrinsic"
fault parameters play a role in the "dynamic plastic evolution" of the fault is defined
by Eq. 4.13 from Piau et al. (2020).

All the following parameters' variations generate fault slip dynamic acceleration
(Table 4.2):

- increase in fluid compressibility above 0.0009 Mpa^{-1},
- an increase of the yield law coefficient α to 35,
- a decrease of fault thickness to 0.2 m associated to an increase in porosity to 0.2.

Figure 4.2b shows the stress path in the case where the fluid compressibility
has increased to 0.002 MPa^{-1}. The "No static plastic" stress evolution triangular
shaped area is delimited by the red line. This area is outside the failure envelop
in the reference case (and thus it does not figure in Fig. 4.2a). As in the reference
case, fault rupture occurs after 1.4 years (point a in Fig. 4.2b). There is then a slow
rupture acceleration until 8.5 years (ab segment in Figs. 4.2b and 4.3a, b). This period
corresponds to a stable plastic phase of dilatant fault deformation characterized by
a total aseismic slip and normal opening displacements respectively of 0.07 and
0.0093 m (Table 4.2). The average slip rate of 0.008 m/year is exceeding the 0.006
m/yr background displacement rate. From b to c, the stress path crosses the "No

Table 4.1 Reference case of a fault at 2.3 km depth

Intact rock		Fault zone		Initial stress	
ρ (kg/m^3)	2500	Thickness (m)	1.2	τ (MPa)	7.7
E (MPa)	5800	Porosity m	0.12	σ'_n (MPa)	16
ν	0.29	M	0.7	Pore pressure	
		P'_0 (MPa)	45	Pf_0 (MPa)	20
		α	21	ΔPf (MPa)	6
		Fluid compressibility		Background displacement rate	
		C_{fl} (MPa^{-1})	0.00045	0.006 m/yr	

Fig. 4.2 Cam-Clay fault activation stress path. (a) Reference case. (b) High 0.002 MPa^{-1} fluid compressibility case. (oa) is the elastic phase ending by fault rupture at a. (ab) is the stable plastic phase driven by fluid pressure increase in the fault. (bc) is the unstable case driven by fault pads inertial forces when the stress path crosses the Nosp area (red line). (cd) is a progressive regain in fault plastic stability, ending in the elastic domain at point (d)

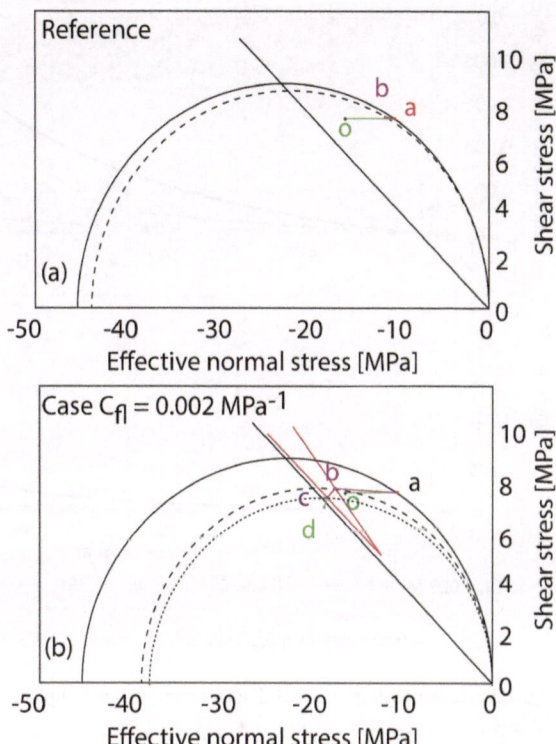

static plastic" stress evolution zone. It corresponds to an unstable and potentially seismic phase characterized by a significant acceleration of slip to a value of 0.007 m/s^2 (or a slip velocity of 0.012 m/s). During the b-c period, a slip amplitude of 0.189 m, larger than the total slip during the aseismic ab period, occurs in about 12 s. It releases a part of the energy stored in the pads, potentially generating seismic waves. Plastic stability is progressively recovered from c to d with deceleration of fault displacements (Fig. 4.2b). Finally, the stress path ends by an elastic phase from d to e (Fig. 4.2b).

Figure 4.3a, b shows that the main factor driving fault to unstable plastic and potentially seismic rupture is the fluid compressibility. The larger the compressibility the shorter is the duration of the stable aseismic slipping period, respectively of 8.4 and 18.1 years for compressibilities of 0.002 and 0.0009 MPa^{-1}. In details, the slip velocity is larger with a higher compressibility but the slip acceleration is smaller (Table 4.2). The model accounts for the change in fluid compressibility through the change in the Skempton coefficient B_s defined as

$$B_s = \frac{\left(1 - \frac{C_{bu}}{C_{bd}}\right)}{\alpha} = \frac{\Delta P_f}{\Delta \sigma_n'}, \tag{4.21}$$

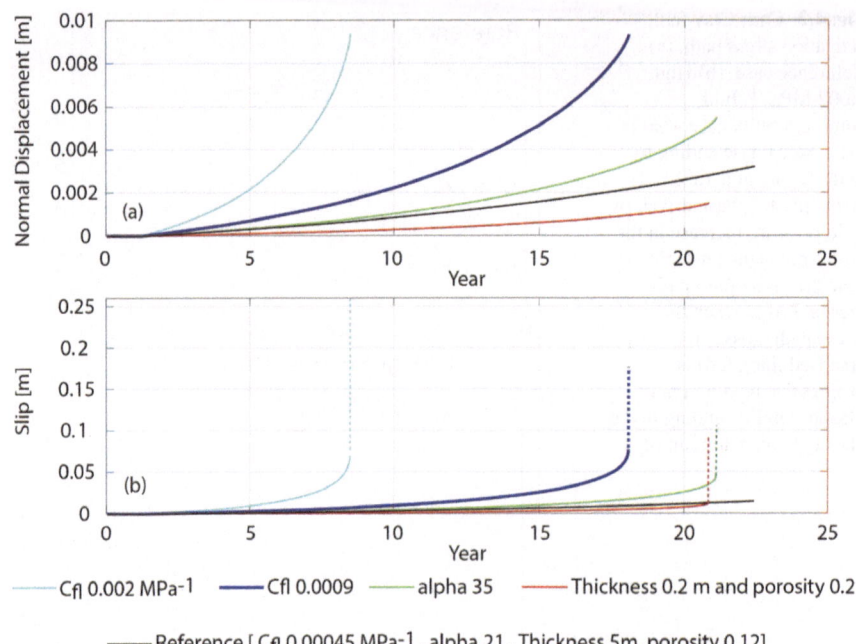

Fig. 4.3 Activated fault normal displacement (a) and slip (b) for the different cases described in Table 4.2

Table 4.2 Summary of fault displacements for different static to dynamic ruptures

Fault properties	Aseismic slip	Seismic slip	Total slip	Normal displacement
Reference C_{fl} 0.00045 MPa^{-1} Thickness 1.2 m Porosity 0.12	0.015 m	0	0.015 m	0.0032 m
C_{fl} 0.0009 MPa^{-1}	0.08 m	0.093 m 0.0026 m/s 0.0013 m/s^2	0.173 m	0.0093 m
C_{fl} 0.002 MPa^{-1}	0.07 m	0.189 m 0.012 m/s 0.007 m/s^2	0.259 m	0.0093 m
$\alpha = 35$	0.045 m	0.066 m 0.0019 m/s 0.0012 m/s^2	0.111 m	0.0055 m
Thickness 0.2 m Porosity 0.2	0.013 m	0.074 m 0.0015 m/s 0.004 m/s^2	0.087 m	0.0015 m

where α is the Biot's coefficient, C_{bu} and C_{bd} the undrained and drained bulk rock compressibility, ΔP_f the variation of fault fluid pressure and $\Delta \sigma_n\prime$ the variation of fault effective normal stress. Note that in Piau et al. (2020) constitutive law, a fault zone Skemton coefficient is defined as S (see Eq. 2.13 in Chap. 2). The undrained rock compressibility is related to the fluid compressibility through the Gassmann's formula:

$$\frac{1}{C_{bu}} = \frac{\emptyset(C_{fl} - C_s) + C_{bd} - C_s}{\emptyset C_{bd}(C_{fl} - C_s) + C_s(C_{bd} - C_s)}. \tag{4.22}$$

For example, if the fluid compressibility C_{fl} increases and becomes much larger than the solid drained compressibility C_{bd}, the above equation resumes to

$$\frac{1}{C_{bu}} \approx \frac{1}{C_{bd}}. \tag{4.23}$$

When supercritical CO_2 mixes with fault initial brine, the Skempton coefficient tends to decrease, and following equation below, the variation in fault effective normal stress tends to equal the variation in total normal stress:

$$\Delta \sigma_n\prime = \Delta \sigma_n - B_s \sigma_n \approx \Delta \sigma_n. \tag{4.24}$$

The increase in the effective normal stress can thus only be related to a change in the fault fluid properties. Figure 4.3 shows that such change can drastically reduce the time to residual strength of the shearing fault zone depending on the fault fluid compressibility increase. In relation to such a faster weakening, the total amount of fault slip and opening is significantly increased. This has a strong effect on fault stability and leakage in the way change of fault fluid properties favors dilation. Nevertheless, the change in fluid properties also reduces the viscosity of the fluid and thus the fluid pressure in the fault should be reduced following the Darcy law. This viscosity effect which is not accounted for in this model may balance the compressibility effect on fault bulk elastic properties.

Note that unstable behavior can occur for a compressibility of 0.0009 MPa^{-1} that corresponds to the estimated compressibility of a brine at the considered CO_2 storage depth. The compressibility of 0.002 MPa^{-1} would figure supercritical CO_2 entering the fault, and replacing the pre-existing brine. These conceptual results highlight that a fluid pressure increase caused by a massive injection of supercritical CO_2 could potentially induce seismicity either by transferring fluid pressure into a given fault zone, and/or by connecting the CO_2 plume to a fault zone.

A large yield law coefficient α and a thin-high porosity fault induce about the same plastic instability that occurs at about 21 years with a potential seismic slip amplitude of about 0.07 m (Table 4.2 and Fig. 4.3a, b). The increase in α favors more aseismic slip and dilation and a larger slip acceleration than the thin-porous fault.

References

Blanpied, M. L., Marone, C. J., Lockner, D. A., Byerlee, J. D., & King, D. P. (1998). Quantitative measure of the variation in fault rheology due to fluid-rock interactions. *Journal of Geophysical Research, 103*(B5), 9691. https://doi.org/10.1029/98JB00162

Bondarenko, N., Podladchikov, Y., & Makhnenko, R. (2022). Hydromechanical impact of basement rock on injection-induced seismicity in Illinois Basin. *Science and Reports, 12*, 15639. https://doi.org/10.1038/s41598-022-19775-4

Bourne, S. J., & Oates, S. J. (2017). Extreme threshold failures within a heterogeneous elastic thin sheet and the spatial-temporal develop-ment of induced seismicity within the Groningen gas field. *Journal of Geophysical Research: Solid Earth, 122*, 10,299–10,32.

Brixel, B., Klepikova, M., Jalali, M.R., Lei, Q., Roques, C., Kriestch, H., & Loew, S. (2020). Tracking fluid flow in shallow crustal fault zones: 1. Insights from single-hole permeability estimates. *Journal of Geophysical Research: Solid Earth, 125*.

Candela, T., Osinga, S., Ampuero, J.-P., Wassing, B., Pluymaekers, M., Fokker, P. A., et al. (2019). Depletion-inducedseismicity at the Groningen gas field: Coulomb rate-and-state models including differential compaction effect. *Journal of Geophysical Research: Solid Earth, 124*, 7081–7104.

Cappa, F., Scuderi, M. M., Collettini, C., Guglielmi, Y., & Avouac, J. P. (2019). Stabilization of fault slip by fluid injection in the laboratory and in situ. *Science Advances*, EAAU4065.

Cebry, S. B. L., Ke, C.-Y., & McLaskey, G. C. (2022). The role of background stress state in fluid-induced aseismic slip and dynamic rupture on a 3-m laboratory fault. *Journal of Geophysical Research: Solid Earth, 127*, e2022JB024371.

Chang, K. W., & Segall, P. (2016). Injection-induced seismicity on basement faults including poro-elastic stressing. *Journal of Geophysical Research: Solid Earth, 121*, 2708–2726. https://doi.org/10.1002/2015JB012561

Cheng, Y., Liu, W., Xu, T., Zhang, Y., Zhang, X., Xing, Y., Feng, B., & Xia, Y. (2023). Seismicity induced by geological CO_2 storage: A review. *Earth-Science Reviews, 239*, 104369.

Collettini, C., Niemeijer, A., Viti, C., Smith, S. A., & Marone, C. (2011). Fault structure, frictional properties and mixed-mode fault slip behavior. *Earth and Planetary Science Letters, 311*(3–4), 316–327. https://doi.org/10.1016/j.epsl.2011.09.020

Cornelio, C., Passelègue, F. X., Spagnuolo, E., Di Toro, G., & Violay, M. (2020). Effect of fluid viscosity on fault reactivation and coseismic weakening. *Journal of Geophysical Research: Solid Earth, 125*, e2019JB018883. https://doi.org/10.1029/2019JB018883

Das, I., & Zoback, M. D. (2013). Long-period long-duration seismic events during hydraulic stimulation of shale and tight-gas reservoirs—Part 2: Location and mechanisms. *Geophysics, 78*(6) (November-December 2013); P. KS97–KS105, 9 FIGS. https://doi.org/10.1190/GEO2013-0165.1

De Barros, L., Cappa, F., Deschamps, A., & Dublanchet, P. (2020). Imbricatedaseismic slip andfluid diffusion drive aseismic swarm in the Corinth Gulf, Greece. *Geophysical Research Letters, 47*, e2020GL087142.

De Barros, L., Cappa, F., Guglielmi, Y., et al. (2019). Energy of injection-induced seismicity predicted from *in-situ* experiments. *Science and Reports, 9*, 4999. https://doi.org/10.1038/s41598-019-41306-x

Dieterich, J. H. (1978). Preseismic fault slip and earthquake prediction. *Journal of Geophysical Research, 83*, 3940–3947.

Dieterich, J. H. (1979). Modeling of rock friction: 1. Experimental results and constitutive equations. *Journal of Geophysical Research, 84*(B5), 2161. https://doi.org/10.1029/JB084iB05p02161

Dieterich, J. (1994). A constitutive law for rate of earthquake production and its application to earthquake clustering. *Journal of Geophysical Research: Solid Earth, 99*(B2), 2601–2618.

Dost, B., Ruigrok, E., & Spetzler, J. (2017). Development of seismicity and probabilistic hazard assessment for the Groningen gas field. *Netherlands Journal of Geosciences, 96*(5), S235–S245. https://doi.org/10.1017/njg.2017.20

Ellsworth, W. L. (2013). Injection-induced earthquakes. *Science, 341*, 1225942. https://doi.org/10.1126/science.1225942

Galgana, G. A., & Hamburger, M. W. (2010). Geodetic observations of active intraplate crustal deformation in the Wabash Valley seismic zone and the Southern Illinois Basin. *Seismological Research Letters, 81*(5), 699–714. https://doi.org/10.1785/gssrl.81.5.699

Geli, L., Piau, J. M., Dziak, R., Maury, V., Fitzenz, D., Coutellier, Q., & Henry, P. (2014). Seismic precursors linked to super-critical fluids at oceanic transform faults. *Nature Geosciences*. https://doi.org/10.1038/NGEO2244

Guglielmi Y., Cappa F., Avouac J. P., Henry P., & Elsworth, D. (2015). Seismicity triggered by fluid-injection-induced aseismic slip. *Science, 348*, 1224. https://doi.org/10.1126/science.aab0476

Guo, H., Brodsky, E. E., Goebel, T. H. W., & Cladouhos, T. T. (2021). Measuring fault zone and host rock hydraulic properties using tidal responses. *Geophysical Research Letters, 48*, e2021GL093986.

Heimisson, E. R., Smith, J. D., Avouac, J. P., & Bourne, S. J. (2022). Coulomb threshold rate-and-state model for fault reactivation: Application to induced seismicity at Groningen. *Geophysical Journal International, 2022*(228), 2061–2072.

Islam, A. W., & Carlson, E. S. (2012). Viscosity models and effects of dissolved CO_2. *GRC Transactions, 36*.

Im, K., Chris Marone, C., & Derek Elsworth, D. (2019). The transition from steady frictional sliding to inertiadominated instability with rate and state friction. *Journal of the Mechanics and Physics of Solids, 122*, 116–125

Ito, Y., & Ikari, M. J. (2015). Velocity- and slip-dependent weakening in simulated fault gouge: Implications for multimode fault slip. *Geophysical Research Letters, 42*(21), 9247–9254. https://doi.org/10.1002/2015GL065829

Jeanne, P., Zhang, Y., & Rutqvist, J. (2020). Influence of hysteretic stress path behavior on seal integrity during gas storage operation in a depleted reservoir. *Journal of Rock Mechanics and Geotechnical Engineering, 12*(4).

Jia, K., et al. (2020). Nonstationary background seismicity rate and evolution of stress changes in the changning salt mining and shale-gas hydraulic fracturing region, Sichuan Basin, China. *Seismological Research Letters, 91*(4), 2170–2181.

Keranen, K. M., Weingarten, M., Abers, G. A., Bekins, B. A., & Ge, S. (2014). Sharp increase in central Oklahoma seismicity since 2008 induced by massive wastewater injection. *Science*. https://doi.org/10.1126/science.1255802

Kolawole, F., Johnston, C. S., Morgan, C. B. et al. (2019). The susceptibility of Oklahoma's basement to seismic reactivation. *National Geoscience, 12*, 839–844 (2019). https://doi.org/10.1038/s41561-019-0440-5

Luu, K., Schoenball, M., Oldenburg, C. M., & Rutqvist, J. (2022). Coupled hydromechanical modeling of induced seismicity from CO_2 injection in the Illinois Basin. *Journal of Geophysical Research: Solid Earth, 127*, e2021JB023496. https://doi.org/10.1029/2021JB023496

Marguin, V., & Simpson, G. (2023). Influence of fluids on earthquakes based on numerical modeling. *Journal of Geophysical Research: Solid Earth, 128*, e2022JB025132. https://doi.org/10.1029/2022JB025132

Marone, C. J. (1998). Laboratory-derived friction laws and their application to seismic faulting. *Annual Review of Earth and Planetary Sciences, 26*(1), 643–696. https://doi.org/10.1146/annurev.earth.26.1.643

Niemeijer, A. R., Boulton, C., Toy, V. G., Townend, J., & Sutherland, R. (2016). Large-displacement, hydrothermal frictional properties of DFDP-1 fault rocks, Alpine Fault, New Zealand: Implications for deep rupture propagation. *Journal of Geophysical Research: Solid Earth, 121*(2), 624–647. https://doi.org/10.1002/2015JB012593

Piau, J. M., Maury, V., & Firenz, D. (2020). Interface plastic constitutive law with end-cap and structural model applied to geological faults behavior. *Journal of Geophysical Research: Solid Earth, 125*.

Rice, J. R., & Ruina, A. L. (1983). Stability of steady frictional slipping. *Journal of Applied Mechanics, 50*(2), 343. https://doi.org/10.1115/1.3167042

Ringrose, P., Andrews, J., Zweigle, P., Furre, A. K., Hern, B., & Bamshad, N. (2022). Why CCS is not like reverse gas engineering. *First Break, 40.*

Ross, Z. E., Cochran, E. S., Trugman, D. T., & Smith, J. D. (2020). 3D fault architecture controls the dynamism of earthquake swarms. *Science, 368,* 1357–1361.

Roy, M., & Marone, C. (1996). Earthquake nucleation on model faults with rate- and state-dependent friction: Effects of inertia. *Journal of Geophysical Research, 101*(86), 13919–13932.

Ruina, A. (1983). Slip instability and state variable friction laws. *Journal of Geophysical Research, 88*(370), 10.

Segall, P., & Lu, S. (2015). Injection-induced seismicity: Poroelastic and earthquake nucleation effects. *Journal of Geophysical Research: Solid Earth, 120,* 5082–5103. https://doi.org/10.1002/2015JB012060

Segall, P., & Rice, J. R. (2015). Dilatancy, compaction, and slip instability of a fluid-infiltrated fault. *Journal of Geophysical Research, 100*(B11), 22155–22171.

Shapiro, S. A. (2015). *Fluid-induced seismicity.* Cambridge University Press.

Sleep, N. H., & Blanpied, M. L. (1994). Ductile creep and compaction: A mechanism for transiently increasing fluid pressure in mostly sealed fault zones. *Pure and Applied Geophysics, 143*(1), 9–40. https://doi.org/10.1007/BF00874322

Snippe, J., Kampman, N., Bisdom, K., Tambach, T., March, R., Maier, C., Phillips, T., Inskip, N. F., Doster, F., & Busch, A. (2022). Modelling of long-term along-fault flow of CO2 from a natural reservoir. *International Journal of Greenhouse Gas Control, 118,* 103666.

Van den Ende, M. P. A., Chen, J., Ampuero, J. P., & Niemeijer, A. R. (2018). A comparison between rate-and-state friction and microphysical models, based on numerical simulations of fault slip. *Tectonophysics, 733,* 273–295.

Verdon, J. P. (2016). Using microseismicity data recorded at the Weyburn CCS-EOR site to assess the likelihood of induced seismic activity. *International Journal of Greenhouse Gas Control, 54,* 421–428.

Vilarrasa, V., Bolster, D., Dentz, M., Olivella, S., & Carrera, J. (2019). Effects of CO_2 compressibility on CO_2 storage in deep saline aquifers. Transport in Porous Media. https://doi.org/10.1007/s11242-010-9582-z

Yang, Y., & Dunham, E. M. (2023). Influence of creep compaction and dilatancy on earthquake sequences and slow slip. *Journal of Geophysical Research: Solid Earth, 128,* e2022JB025969. https://doi.org/10.1029/2022JB025969

Zhai, G., Shirzaei, M., Manga, M., & Chen, X. (2019). Pore pressure diffusion, enhanced by poroelastic stresses, controls induced seismicity in Oklahoma. *PNAS, 116*(33), 16228–16233.

Zhang, Y., Person, M., Rupp, J., Ellett, K., Celia, M.A., Gable, C. W., Bowen, B., Evans, J., Bandilla, K., Mozley, P., Dewers, T., & Elliot, T. (2013). *Hydrogeologic controls on induced seismicity in crystalline basement rocks due to fluid injection into basal reservoirs* (Vol. 51, No. 4, pp. 525–538)–Groundwater–July-August 2013.

Zhou, Q., Birkholzer, J. T., Mehnert, E., Lin, Y. F., & Zhang, K. (2010). Modeling basin- and plume-scale processes of CO_2 storage for full-scale deployment. *Ground Water, 48*(4), 494–514. https://doi.org/10.1111/j.1745-6584.2009.00657

Zhu, W., Allison, K. L., Dunham, E. M., et al. (2020). Fault valving and pore pressure evolution in simulations of earthquake sequences and aseismic slip. *Nature Communications, 11,* 4833. https://doi.org/10.1038/s41467-020-18598-z

Chapter 5
Conclusion—CO$_2$ Storage Scenarios and Basin Scale Implications

Abstract At first glance, fluid leakage through fault zones and induced seismicity may look like different issues dealing with the different risks that they put on the management of a CO$_2$ storage project in a deep reservoir. Leakage risk concerns the loss of caprock integrity caused by CO$_2$ penetrating faults cutting through the reservoir and the caprock, thus located at and above the injection reservoir depth. Seismicity is in general localized deeper on faults, mostly disconnected from the injection reservoir and rooted into the upper basement rocks. Nevertheless, this review highlights that workflows integrating fluid leakage and induced seismicity together should be more beneficial to the management of CO$_2$ storage projects, first because observations show earthquakes associated to leakage in caprocks and downward pressurization of faults favors induced seismicity in the basement.

Keyword Fault zone · Creep · Permeability · Seismicity · Cam-Clay constitutive laws · Hub of CO$_2$ projects · Basin scale

Observations have shown that in situ faults hydromechanical behavior strongly depends on confining and deviatoric stresses, and on the ratio between bulk strain versus localized slip. We suggest that most of the fault permeability change during activation occurs at low cumulated slip, when the bulk fault zone dilation under plastic shear is dominant. Three orders of magnitude fault permeability increases are possible at least locally. This is a relatively slow process associated with progressive fault zone weakening driven by poroelastic and pressure diffusion rates enhanced since the start of an injection project. A slow fault creep acceleration can occur spanning over months-to-tens of years depending on fault hydromechanical and frictional properties. Large, cumulated slip does not scale with a large fault permeability increase. This is explained by a localization of the slip on a frictional interface or thin band within the fault zone. If this is only partially altering permeability change, it can lead to slip acceleration and seismicity. At a basin scale context, we identify that 0.01-to-0.08 m slip can then be triggered over 8-to-10 s durations that could correspond to moderate magnitude earthquakes. This means that faults must be described as zones and not only as simple zero-thickness frictional dilatant or contracting interfaces. In addition, more general end-cap yield criterions are required to consider that the same fault zone be ductile and brittle depending on the fault properties, the variability of stress paths leading to fault activation and on the changes in fault fluid properties induced by a storage project injecting supercritical CO$_2$ into a brine. We suggest that the evolution between bulk plastic strain and frictional slip is critical to better assess the magnitude and the rate of fault permeability and slip variations. One possible avenue would be to adapt the Cam-Clay constitutive laws to describe the hydromechanical behavior of fault zones at basin scale. Tested here, this physics proves to be more general than a frictional physics based on Coulomb failure and rate-and-state normalism, because (1) it allows considering that evolution from bulk plasticity to frictional slip including the coupling with permeability change, (2) it provides an end-cap yield criterion that allows to describe both dilatant softening or contraction strengthening of an activated fault zone depending on the stress scenario where the Coulomb yield surface is limited to brittle behavior and has no end-cap to describe ductility and (3) it seems possible to describe fault transition to seismicity without a priori compared to the rate-and-state approach that requires to predefine whether or not the fault will rupture seismically through pre-setting (a-b) parameter negative.

Furthermore, the deployment of a hub of CO$_2$ storage projects will change effective stresses and pore fluid properties in a significant volume of the basin that is way larger than the injected reservoir, potentially bringing many faults to failure. This review shows that failure can start slowly and progressively accelerate to the seismic instability. The consequence is that fault failure may remain undetected until the first earthquake. What may happen is that a slow failure period takes years while other CO$_2$ storage projects are deployed in the same basin, eventually in the same reservoir. When an earthquake happens and although it results from a long-term fault slip acceleration initiated by the first project, it is difficult to clearly relate it to this first project. This can be particularly problematic for earthquakes localized relatively far from the injection wells, and on faults clearly disconnected to the injection reservoir. The second consequence of slow failure is that fault permeability may drastically change at the beginning of it, again being hardly detectable with conventional seismic. In addition, it does not seem that permeability will decrease so much with increasing cumulated slip thus over years of slow failure. Finally, even if slip rate decreases, the changes at basin scale may maintain background rates that exceed the fault sealing rates in many cases. Though this "dark" picture must be moderated by the strong basin-dependent variability of fault population geological characteristics. Nevertheless, we suggest that the effects of faults slow movements should be better integrated in geomechanical models and that monitoring techniques be developed to better capture strain rates and fault displacement rates at different depths of a basin. In this way, distributed optical fibers cemented between well-casing coupled with very high-resolution borehole displacement tools deployed inside casing offer promising perspectives. Supercritical CO$_2$ replacing fault pore fluid is another key concern. There are very few data on the role of fluid properties on fault rupture, and a drastic lack of field scale experiments. This review "only" shows a part of physics that concerns the effect of the change of fluid compressibility in accelerating fault dilatant weakening. Nevertheless, it shows that a slight change may be enough to cause a drastic slip acceleration. Here again, there is too little information for a better view of the key physics since we can also consider that if the leakage flow path propagates fast up dip along the fault, CO$_2$ may help evacuate pressure and eventually tend to relatively stabilize the fault.

Perspectives are multiple if we run to a more general physics that could apprehend many fault leakage and induced seismicity scenarios. First is about observations needed to calibrate such models. As mentioned above, methods to accurately monitor micro-to-millimetric strain and displacement rates at relevant depths need to be considered. The way to estimate in situ bulk fault zone strength and to monitor its long-term evolution is also needed. Overall, there is a need to intensify development of methods that capture long-term processes at fault zone and basin scales. The second point is may be more fundamental but the conversion of seismic slip into earthquake magnitudes is poorly constrained in the field of a multilayered system in a large basin. Even being able to calculate a dynamic slip, there still is the difficult question to convert into earthquake source properties that may look very different in layered basins compared to what is studied in the deeper seismogenic zone. The third point is the interaction between storage projects and how they can modify fault stability. Very few studies anticipate this key question that will arise in the very near future.

Correction to: A Review of CO_2 Storage Integrity and Fault Zone Risk

Correction to:
Y.Guglielmi, *A Review of CO_2 Storage Integrity and Fault Zone Risk*, **SpringerBriefs in Earth System Sciences, https://doi.org/10.1007/978-3-031-81529-4**

The original version of the book has been updated with the last few lines under the Acknowledgements in the FM. The book has been updated with the changes.

The updated version of this book can be found at
https://doi.org/10.1007/978-3-031-81529-4

C1
Y. Guglielmi, *A Review of CO_2 Storage Integrity and Fault Zone Risk*,
SpringerBriefs in Earth System Sciences, https://doi.org/10.1007/978-3-031-81529-4_6